AQA Science

Exclusively endorsed and approved by AQA

Revision Guide

Pauline Anning • Nigel English • John Scottow

Series Editor: Lawrie Ryan

GCSE Additional Science

Published in 2006 by:
Nelson Thornes Ltd
Delta Place
27 Bath Road
CHELTENHAM
GL53 7TH
United Kingdom

09 10 / 10 9 8 7 6 5

A catalogue record for this book is available from the British Library

ISBN 9780 7487 8311 3

Cover photographs: embryo by Biophoto/Science Photo Library; chemical crystals by Photodisc 4 (NT); static electricity by Photodisc 29 (NT)

Cover bubble illustration by Andy Parker

Illustrations by Bede Illustration, Kevin Jones Associates and Roger Penwill

Page make-up by Wearset Ltd

Printed in China by 1010 Printing International Ltd

Acknowledgements

Alamy/Popperfoto 71br, 81bl; **Axon Images** 1mr, 9l; **Corel 637 (NT)** 87tr; **Corel 710 (NT)** 56br; **Digital Vision 1 (NT)** 59t; **Jim Breithaupt** 98bl; **Nigel English** 14tr; **Photodisc 10 (NT)** 23tr; **Photodisc 19 (NT)** 1br; **Photodisc 45 (NT)** 21tl, 22br; **Photodisc 71 (NT)** 22bm, 23tm; **Rover Group** 90m; **Science Photo Library/Andrew Lambert Photography** 65tl, 87m, 97l, /**CNRI** 28bl, /**Cordelia Molloy** 6b, 57tl, /**Eye Of Science** 17br, /**J. C. Revy** 3m, 17tr, /**Keith Kent** 71tl, /**Mark Clarke** 17ml, /**Steve Gschmeissner** 3mr, /**TRL LTD.** 71mr, **Transport & Road Research Laboratory** 83tr

Picture research by Stuart Sweatmore, Science Photo Library and johnbailey@ntworld.com.

Every effort has been made to trace all the copyright holders, but if any have been overlooked the publisher will be pleased to make the necessary arrangements at the first opportunity.

The following people have made an invaluable contribution to this book:

Jim Breithaupt, Ann Fullick, Patrick Fullick, Paul Lister, Niva Miles.

How to answer questions

Question speak

Command word or phrase	What am I being asked to do?
compare	State the similarities and the differences between two or more things.
complete	Write words or numbers in the gaps provided.
describe	Use words and/or diagrams to say how something looks or how something happens.
describe, as fully as you can	There will be more than one mark for the question so make sure you write the answer in detail.
draw	Make a drawing to show the important features of something.
draw a bar chart / graph	Use given data to draw a bar chart or plot a graph. For a graph, draw a line of best fit.
explain	Apply reasoning to account for the way something is or why something has happened. It is not enough to list reasons without discussing their relevance.
give / name / state	This only needs a short answer without explanation.
list	Write the information asked for in the form of a list.
predict	Say what you think will happen based on your knowledge and using information you may be given.
sketch	A sketch requires less detail than a drawing but should be clear and concise. A sketch graph does not have to be drawn to scale but it should be the appropriate shape and have labelled axes.
suggest	There may be a variety of acceptable answers rather than one single answer. You may need to apply scientific knowledge and/or principles in an unfamiliar context.
use the information	Your answer **must** be based on information given in some form within the question.
what is meant by	You need to give a definition. You may also need to add some relevant comments.

Diagrams

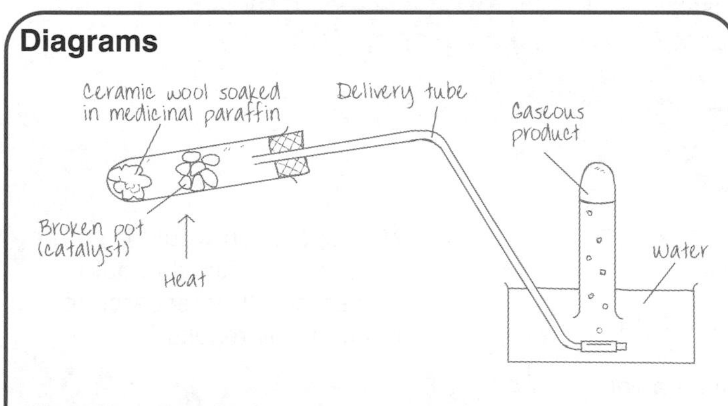

Things to remember:

- Draw diagrams in pencil.
- The diagram needs to be large enough to see any important details.
- Light colouring could be used to improve clarity.
- The diagram should be fully labelled.
- Label lines should be thin and end at the point on the diagram that corresponds to the label.

How long should my answer be?

Things to consider:

1 How many lines have been given for the answer?
- One line suggests a single word or sentence. Several lines suggest a longer and more detailed answer is needed.

2 How many marks is the answer worth?
- There is usually one mark for each valid point. So for example, to get all of the marks available for a three mark question you will have to make three different, valid points.

3 As well as lines, is there also a blank space?
- Does the question require you to draw a diagram as part of your answer?
- You may have the option to draw a diagram as part of your answer.

Graphs

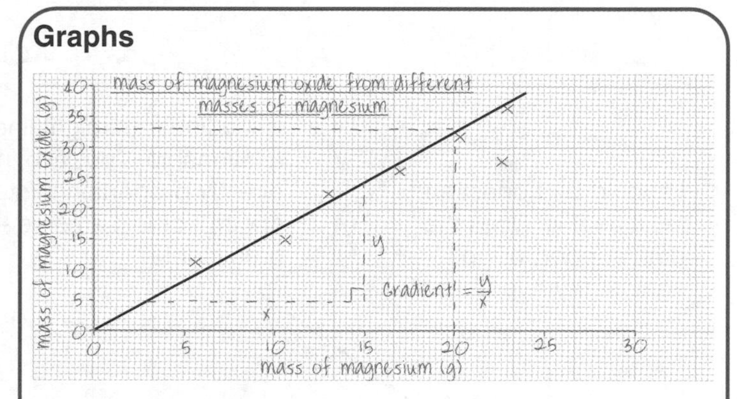

Things to remember:

- Choose sensible scales so the graph takes up most of the grid.
- Don't choose scales that will leave small squares equal to 3 as it is difficult to plot values with sufficient accuracy on such scales.
- Label the axes including units.
- Plot all points accurately by drawing small crosses using a fine pencil.
- Don't try to draw a line through every point. Draw a line of best fit.
- A line of best fit does not have to go through the origin.
- When drawing a line of best fit, don't include any points which obviously don't fit the pattern.
- The graph should have a title stating what it is.
- To find a corresponding value on the y-axis, draw a vertical line from the x-axis to the line on the graph, and a horizontal line across to the y-axis. Find a corresponding value on the x-axis in a similar way.
- The gradient or slope of a line on a graph is the amount it changes on the y-axis divided by the amount it changes on the x-axis. (See the graph above.)

Calculations

- Write down the equation you are going to use, if it is not already given.
- If you need to, rearrange the equation.
- Make sure that the quantities you put into the equation are in the right units. For example you may need to change centimetres to metres or grams to kilograms.
- Show the stages in your working. Even if your answer is wrong you can still gain method marks.
- If you have used a calculator make sure that your answer makes sense. Try doing the calculation in your head with rounded numbers.
- Give a unit with your final answer, if one is not already given.
- Be neat. Write numbers clearly. If the examiner cannot read what you have written your work will not gain credit. It may help to write a few words to explain what you have done.

How to use the 'How Science Works' snake

The snake brings together all of those ideas that you have learned about 'How Science Works'. You can join the snake at different places – an investigation might start an observation, testing might start at trial run.

How do you think you could use the snake on how marble statues wear away with acid rain? Try working through the snake using this example – then try it on other work you've carried out in class.

Remember there really is no end to the snake – when you reach the tail it is time for fresh observations. Science always builds on itself – theories are constantly improving.

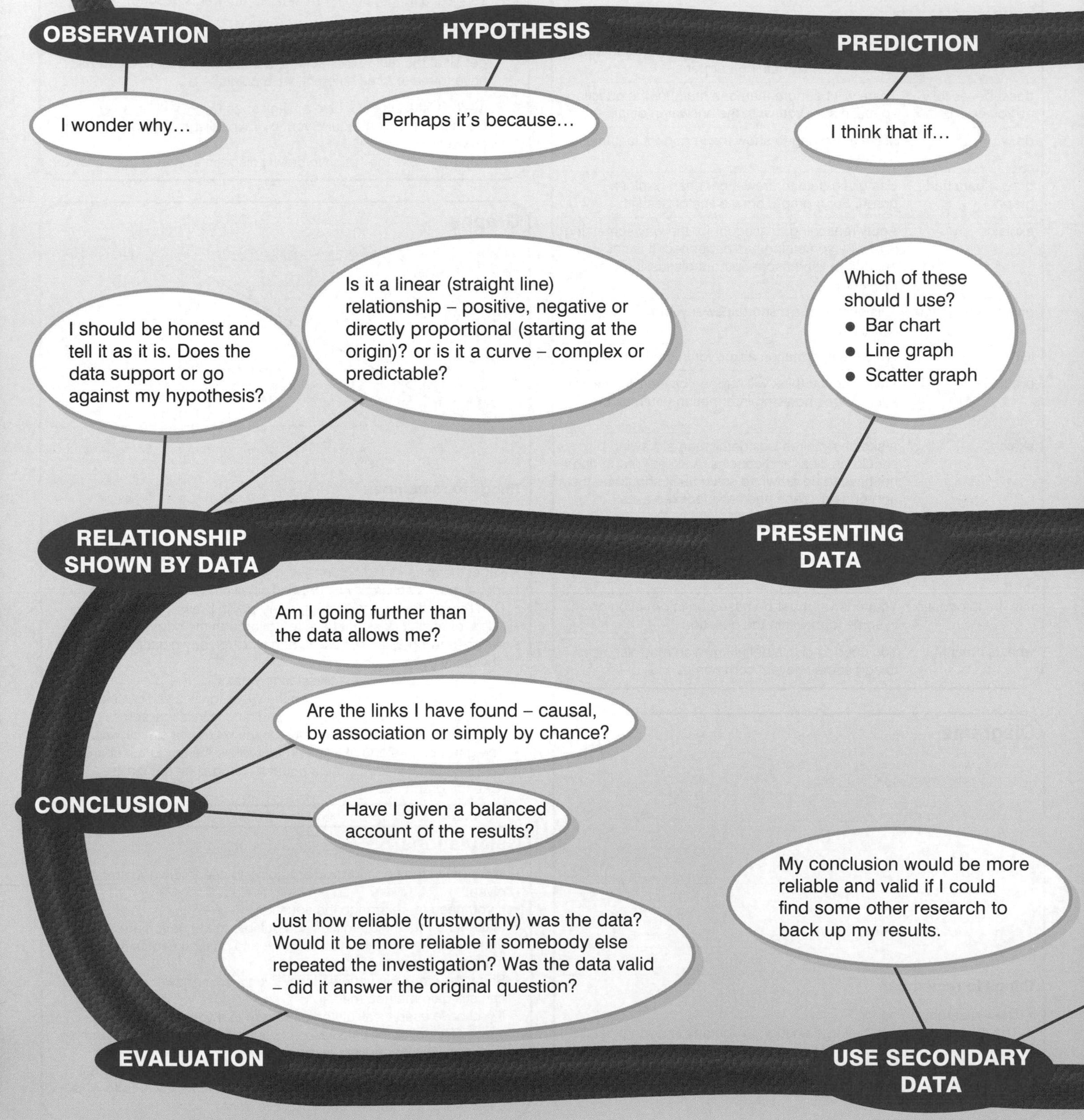

There are still many questions that we cannot answer in science.

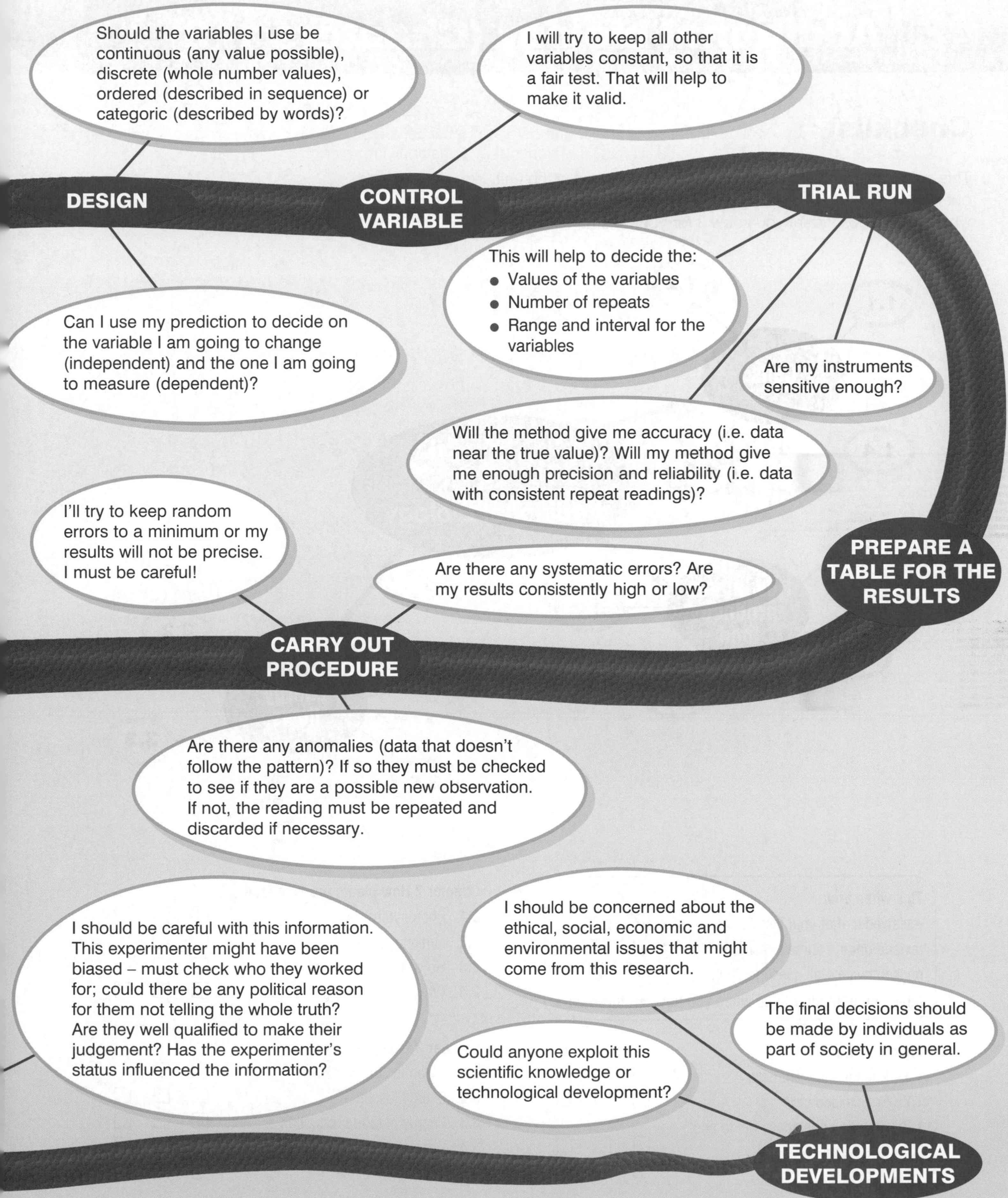

Should the variables I use be continuous (any value possible), discrete (whole number values), ordered (described in sequence) or categoric (described by words)?

I will try to keep all other variables constant, so that it is a fair test. That will help to make it valid.

DESIGN

CONTROL VARIABLE

TRIAL RUN

This will help to decide the:

- Values of the variables
- Number of repeats
- Range and interval for the variables

Can I use my prediction to decide on the variable I am going to change (independent) and the one I am going to measure (dependent)?

Are my instruments sensitive enough?

Will the method give me accuracy (i.e. data near the true value)? Will my method give me enough precision and reliability (i.e. data with consistent repeat readings)?

I'll try to keep random errors to a minimum or my results will not be precise. I must be careful!

Are there any systematic errors? Are my results consistently high or low?

PREPARE A TABLE FOR THE RESULTS

CARRY OUT PROCEDURE

Are there any anomalies (data that doesn't follow the pattern)? If so they must be checked to see if they are a possible new observation. If not, the reading must be repeated and discarded if necessary.

I should be careful with this information. This experimenter might have been biased – must check who they worked for; could there be any political reason for them not telling the whole truth? Are they well qualified to make their judgement? Has the experimenter's status influenced the information?

I should be concerned about the ethical, social, economic and environmental issues that might come from this research.

The final decisions should be made by individuals as part of society in general.

Could anyone exploit this scientific knowledge or technological development?

TECHNOLOGICAL DEVELOPMENTS

There are questions that science cannot answer at all – such as 'Should we…?' questions.

B2 | Additional biology (Chapters 1–3)

Checklist

This spider diagram shows the topics in the first half of the unit. You can copy it out and add your notes and questions around it, or cross off each section when you feel confident you know it for your exams.

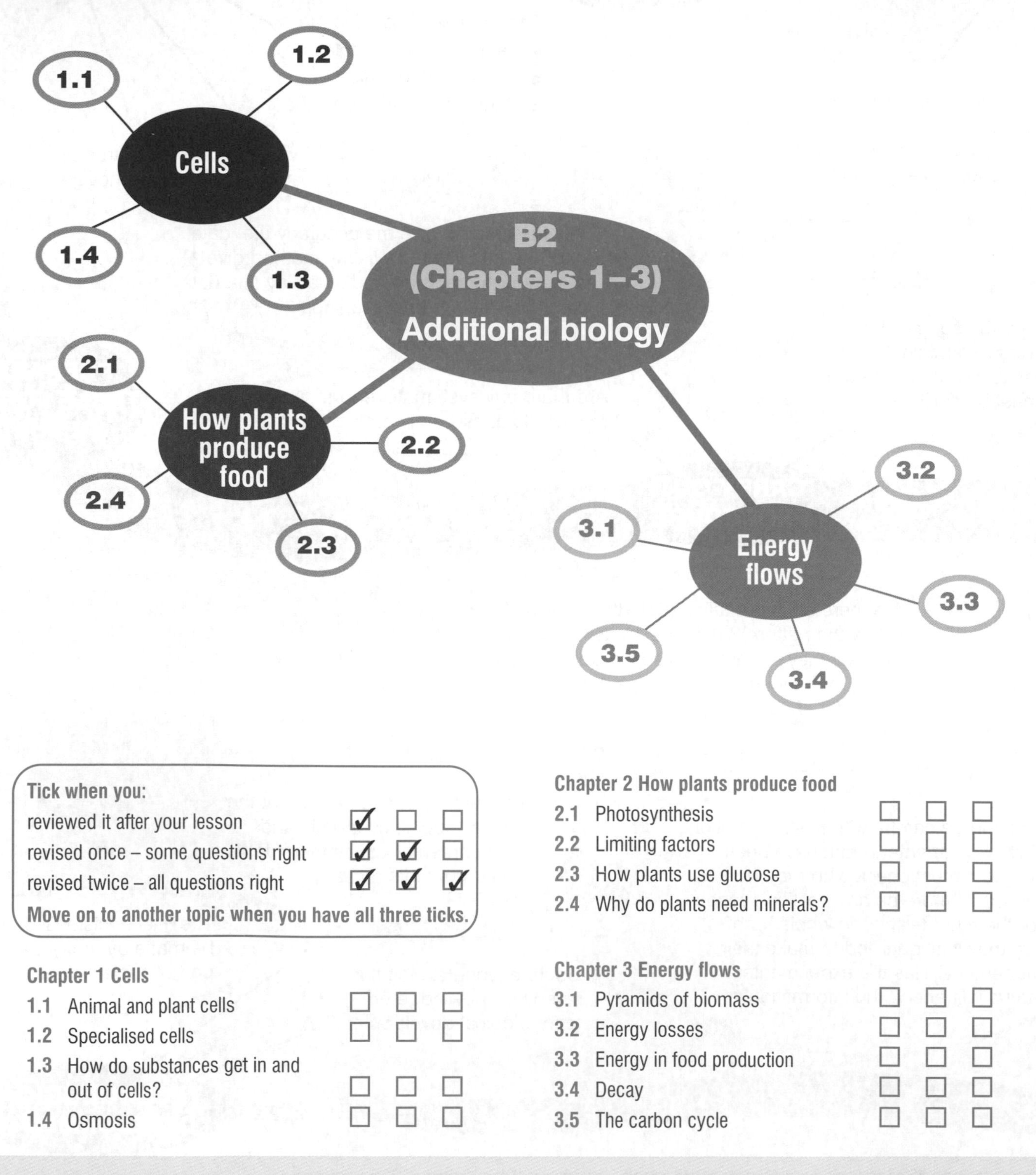

Tick when you:

reviewed it after your lesson	✓	☐	☐
revised once – some questions right	✓	✓	☐
revised twice – all questions right	✓	✓	✓

Move on to another topic when you have all three ticks.

Chapter 1 Cells

1.1 Animal and plant cells	☐	☐	☐
1.2 Specialised cells	☐	☐	☐
1.3 How do substances get in and out of cells?	☐	☐	☐
1.4 Osmosis	☐	☐	☐

Chapter 2 How plants produce food

2.1 Photosynthesis	☐	☐	☐
2.2 Limiting factors	☐	☐	☐
2.3 How plants use glucose	☐	☐	☐
2.4 Why do plants need minerals?	☐	☐	☐

Chapter 3 Energy flows

3.1 Pyramids of biomass	☐	☐	☐
3.2 Energy losses	☐	☐	☐
3.3 Energy in food production	☐	☐	☐
3.4 Decay	☐	☐	☐
3.5 The carbon cycle	☐	☐	☐

What are you expected to know?

Chapter 1 Cells See students' book pages 4–15

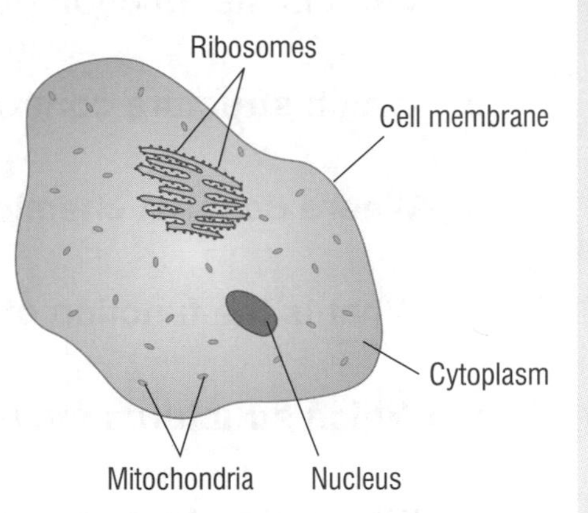

- Animal and plant cells contain a nucleus, cell membrane, cytoplasm, mitochondria and ribosomes.
- The functions (jobs) of these different organelles.
- Plant cells also have a cell wall, chloroplasts and a permanent vacuole.
- The functions of these organelles in plants.
- Some cells are specialised to carry out their function.
- Materials are transported across cell membranes by diffusion and osmosis.
- How the processes of diffusion and osmosis are similar and what the differences are between them.

Chapter 2 How plants produce food

See students' book pages 16–27

- Plants produce carbohydrates through photosynthesis and some of these are stored.
- The factors that limit the rate of photosynthesis are the:
  - level of carbon dioxide
  - temperature
  - light intensity.
- Plants need nitrate and magnesium in order to grow properly.
- If a plant is deficient in these nutrients then the plant will show symptoms.

Chapter 3 Energy flows See students' book pages 28–41

- A pyramid of biomass tells you the mass of the different organisms in a food chain (and it can give you a better picture of a food chain than a pyramid of numbers).
- Energy is transferred in a food chain from one organism to the next.
- Energy is lost in food chains.
- The shorter a food chain is the more efficient food production will be.
- We can artificially improve the efficiency of a food chain when we are producing food, but this can be controversial.
- Materials are recycled by microorganisms.
- A stable community recycles all of the nutrients.
- The carbon cycle.

B2 1 Pre Test: Cells

1. What is the function (job) of the nucleus?
2. Which structure controls the movement into and out of a cell?
3. Where do most chemical reactions take place in a cell?
4. What is the function of ribosomes?
5. Which structures are necessary for photosynthesis to take place?
6. Which structures are necessary for aerobic respiration to take place?
7. Why are some plant and animal cells specialised?
8. How can diffusion be defined?
9. What substance(s) might diffuse into or out of cells?
10. How is osmosis different to diffusion?

students' book page 4

B2 1.1 Animal and plant cells

KEY POINTS

1. Animal and plant cells have structures that enable them to do their jobs.
2. Plant cells have some structures which animal cells don't have.

AQA EXAMINER SAYS...

Remember that the cell membrane can control the movement of materials into and out of the cell. The cell wall, in plants, does not do this. It is there for support.

Animal and plant cells have some structures in common; they have:

- a nucleus to control the cell's activities
- cytoplasm where many chemical reactions take place
- a cell membrane that controls the movement of materials
- mitochondria where energy is released during aerobic respiration
- ribosomes where proteins are made (synthesised).

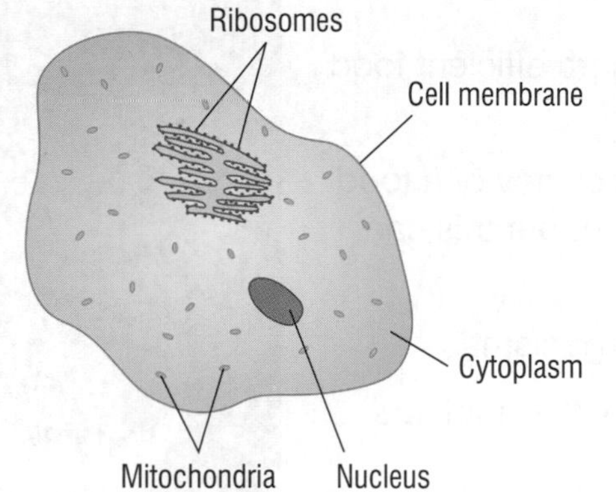

A simple **animal cell** like this shows the features that are common to all living cells

Plant cells also have:

- a rigid cell wall for support
- chloroplasts that contain chlorophyll for photosynthesis
- a permanent vacuole containing cell sap.

Key words: nucleus, cytoplasm, cell membrane, mitochondria, ribosomes, cell wall, chloroplasts, vacuole

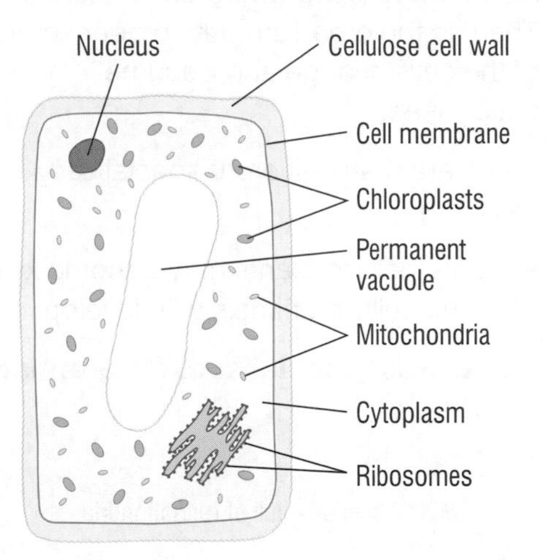

A **plant cell** has many features in common with an animal cell, but others which are unique to plants

GET IT RIGHT!

Make sure you can label an animal cell and a plant cell and know the function of each of their parts.

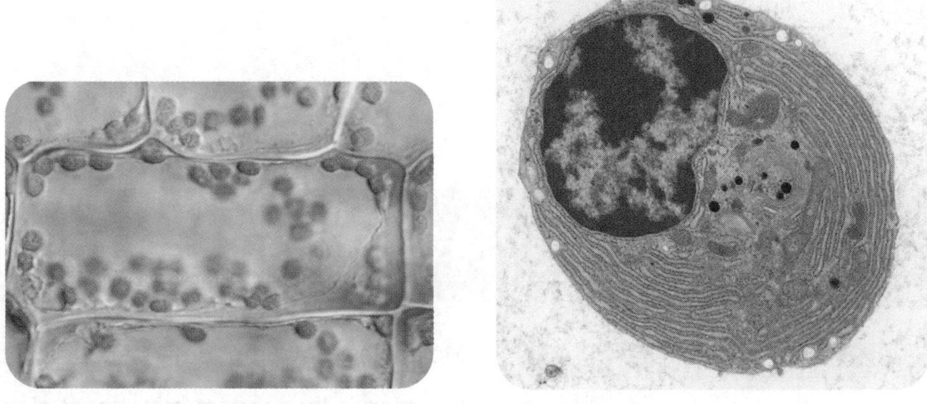

Diagrams of cells are much easier to understand than the real thing seen under a microscope. These pictures show a magnified plant cell and animal cell.

EXAM HINT

Remember that not all plant cells have chloroplasts . . . and don't confuse chloroplasts with chlorophyll.

CHECK YOURSELF

1 Name three structures common to both plant and animal cells.

2 Where does aerobic respiration take place?

3 Name two materials that move across the cell membrane.

students' book page 6

B2 1.2 Specialised cells

KEY POINTS

1. As organisms develop, some of their cells become specialised to carry out particular jobs. This is called 'differentiation'.
2. Differentiation happens much earlier in the development of animals than it does in plants.

When an egg is fertilised it begins to grow and develop.

At first there is a growing ball of cells. Then as the organism gets bigger some of the cells change and become specialised.

There are many different specialised cells, e.g.

- Some cells in plants may become xylem or root hair cells.
- Some cells in animals will develop into nerve or sperm cells.

Key words: specialised, differentiation

EXAM HINTS

In an exam it is not only important to remember that there are specialised cells. It is just as important to remember how their structure makes them suitable to carry out the job that they do.

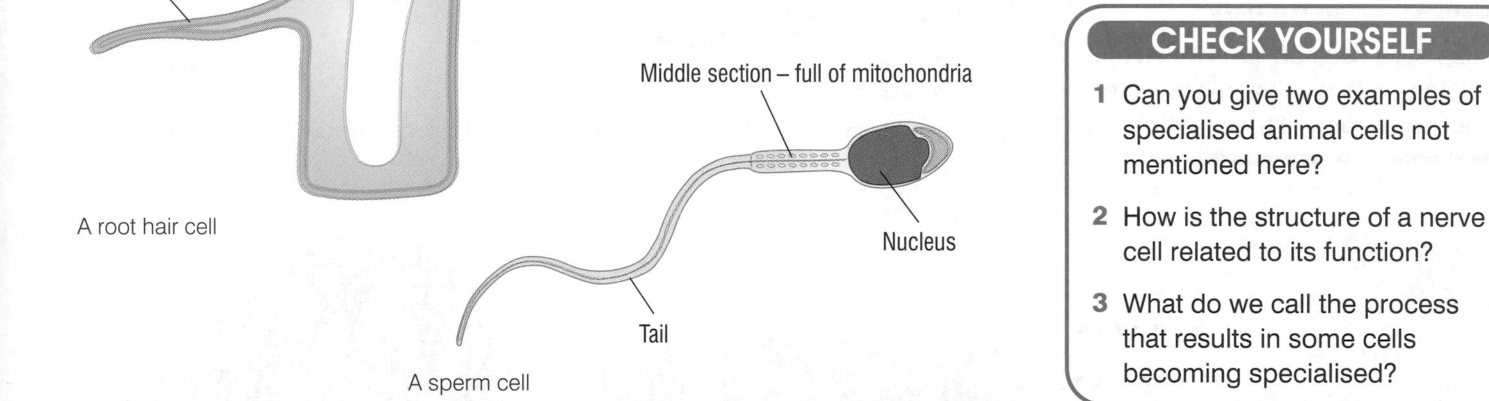

A root hair cell

A sperm cell

CHECK YOURSELF

1. Can you give two examples of specialised animal cells not mentioned here?
2. How is the structure of a nerve cell related to its function?
3. What do we call the process that results in some cells becoming specialised?

students' book page 8

B2 1.3 How do substances get in and out of cells?

KEY POINT

Diffusion is the result of random movement. It does not require any energy from the cell.

Molecules move randomly because of the energy they have.

Diffusion is the random movement of molecules from an area of *high concentration* to an area of *lower concentration.*

The larger the difference in concentration, the faster the rate of diffusion. Examples are:

- the diffusion of oxygen into the cells of the body from the blood stream as the cells are respiring (and using up oxygen)
- the diffusion of carbon dioxide into actively photosynthesising plant cells
- the diffusion of simple sugars and amino acids from the gut through cell membranes.

Key words: random, diffusion, concentration

GET IT RIGHT!

Remember that diffusion does not require energy from the cells. It is simply the random movement of particles. The greater the difference in concentrations, the faster diffusion takes place.

CHECK YOURSELF

1. Why do particles move randomly?
2. Can you think of another example of diffusion in living cells?
3. Why is diffusion important?

students' book page 10

B2 1.4 Osmosis

KEY POINT

Osmosis is a special case of diffusion involving a partially (or semi-) permeable membrane.

AQA EXAMINER SAYS...

Remember osmosis is just a special case of diffusion. It only involves the movement of water and it takes place across a partially permeable membrane.

Osmosis is the movement of water.

Just like diffusion, the movement of molecules is random and requires no energy from the cell.

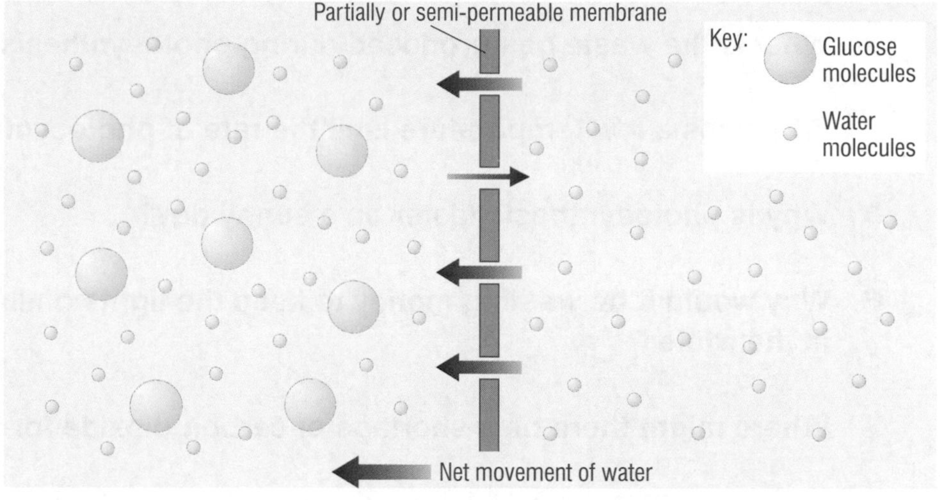

Osmosis

CHECK YOURSELF

1 How is osmosis like diffusion?

2 What do you think 'partially permeable' means?

3 Why do cells need water?

Osmosis is the diffusion of water across a partially permeable membrane from a dilute solution to a more concentrated solution. No solute molecules can move across the membrane. The cell membrane is partially permeable.

Water is needed to support cells and because chemical reactions take place in solution.

Key words: osmosis, partially permeable, solute

B2 1 End of chapter questions

1 **What is the function of ribosomes?**

2 **Suggest how a sperm cell is adapted to its function.**

3 **Why does diffusion require no energy from the cell?**

4 **Why would it be incorrect to say that in osmosis water only passes from the weaker to the stronger solution?**

5 **What substance, necessary for photosynthesis, do chloroplasts contain?**

6 **What is the function of the cell wall in plants?**

7 **Why does oxygen in the blood diffuse into respiring cells?**

8 **Which part of a cell's structure is 'partially permeable'?**

B2 2 Pre Test: How plants produce food

1. Where does the energy for photosynthesis come from?
2. Which carbohydrate is produced during photosynthesis?
3. What is the waste gas produced during photosynthesis?
4. Why does a low temperature limit the rate of photosynthesis?
5. Why is photosynthesis faster on a sunny day?
6. Why would it be wasting money to keep the lights on in a very cold greenhouse in the winter?
7. Where might there be a shortage of carbon dioxide for photosynthesis?
8. Why do plants need nitrate?
9. What would a plant that is short of nitrate look like?
10. Why do plants short of magnesium have yellow leaves?

students' book page 16

B2 2.1 Photosynthesis

KEY POINTS

1. Photosynthesis can only be carried out by green plants.
2. Chlorophyll traps the Sun's energy.

GET IT RIGHT!

Remember that chlorophyll *traps* the Sun's energy, it does not produce any energy of its own.

The equation for photosynthesis is:

carbon dioxide + water **(+ light energy)** → glucose + oxygen

The carbon dioxide is taken in by the leaves, and the water by the roots.

The chlorophyll traps the energy needed for photosynthesis.

In photosynthesis the sugar glucose (a carbohydrate) is made. Oxygen is given off as a waste gas.

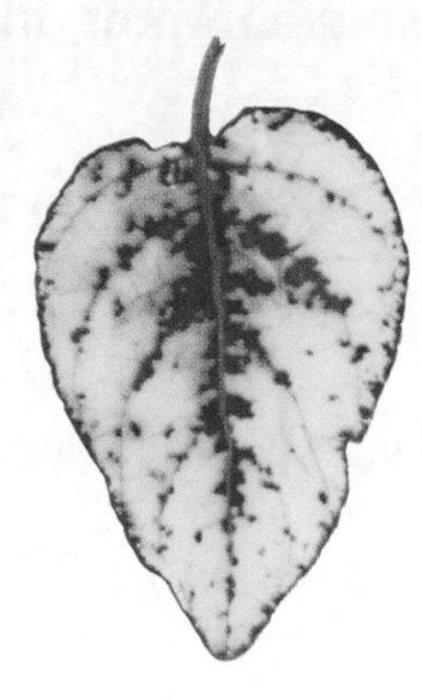

These leaves came from a plant which had been kept in the light for several hours. Only the green parts of the leaf made their own starch which turns the iodine solution blue-black.

The structure of a leaf

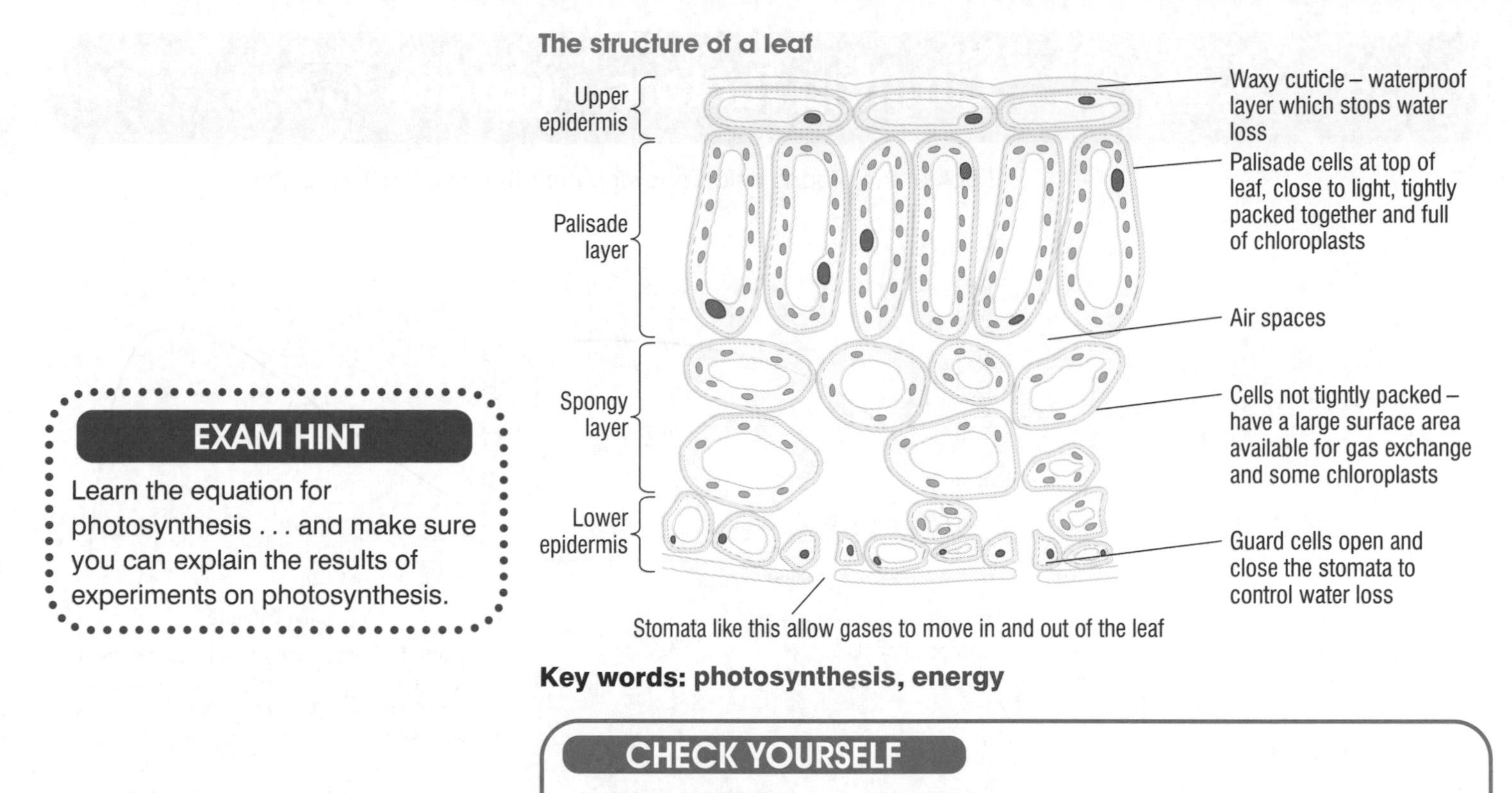

Key words: photosynthesis, energy

CHECK YOURSELF

1 Where does the energy for photosynthesis come from?

2 What are the two substances necessary to make the glucose?

3 Why is the waste product of photosynthesis so important?

students' book page 18

B2 2.2 Limiting factors

KEY POINT

If certain things are in short supply, they will slow down the rate of photosynthesis. Plant growers need to know this, otherwise they could waste money.

A lack of light would slow down the rate of photosynthesis as light provides the energy for the process. Chlorophyll traps the light. Even on sunny days, light may be limited on the floor of a wood or rain forest.

If it is cold, then enzymes do not work effectively and this will slow down the rate.

If there is too little carbon dioxide, then the rate will slow down. Carbon dioxide may be limited in an enclosed space, e.g. in a greenhouse on a sunny day or in a rapidly photosynthesising rain forest.

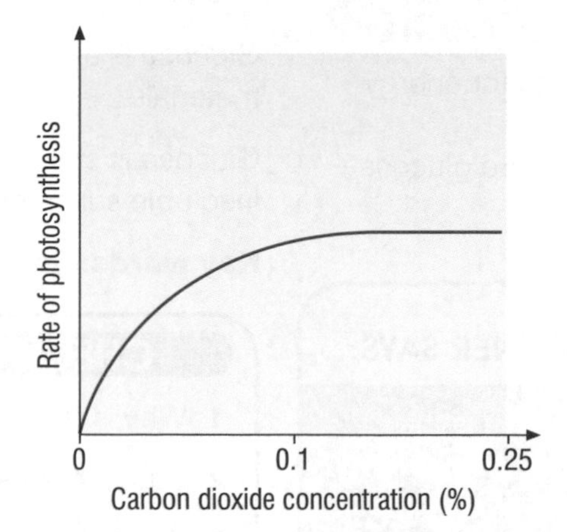

This graph shows the effect of increasing carbon dioxide levels on the rate of photosynthesis at a particular light level and temperature. Eventually one of the other factors becomes limiting.

students' book
page 18

B2 2.2 Limiting factors (cont.)

Look at the graphs below showing the other two limiting factors:

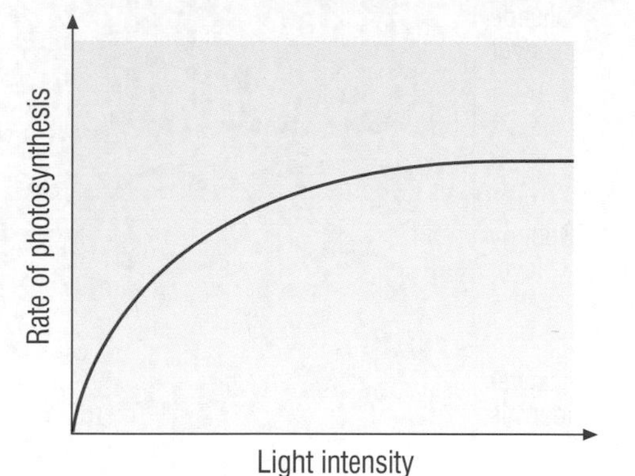

The effect of light intensity on the rate of photosynthesis

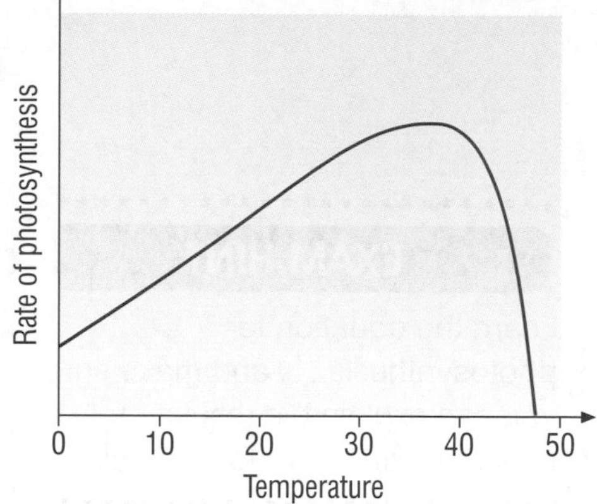

The rate of photosynthesis increases steadily with a rise in temperature up to a certain point. After this the enzymes are destroyed and the reaction stops completely.

For anyone growing plants, there is no point increasing one of these factors if the photosynthesis is still limited by another one. They would just be wasting money.

Key words: limiting, rate

AQA EXAMINER SAYS...

In an exam it may be quite easy to remember these three limiting factors. However you may have to explain why each limits the rate. The most difficult questions involve you interpreting graphs. Try to make sure you practise this skill.

CHECK YOURSELF

1 Why does photosynthesis slow down on a cold day?

2 Why would there be little point heating a greenhouse on a summer's day?

3 Other than in an enclosed space (e.g. a greenhouse) where might photosynthesis be limited by the amount of CO_2?

students' book
page 20

B2 2.3 How plants use glucose

KEY POINTS

1 Plants use glucose for energy (respiration).
2 Most plants can store glucose as starch.

The product of photosynthesis is glucose. Glucose is used for respiration.

Glucose is also combined with other nutrients by the plant to produce new materials.

Glucose is stored, by some plants, as insoluble starch. It is stored as an insoluble substance so that it has no effect on osmosis.

Key words: glucose, starch, insoluble

Remember that glucose is the substrate for respiration. In other words it contains the energy that is released through respiration.

CHECK YOURSELF

1 Why is glucose stored as an insoluble compound?

2 Glucose is used for respiration. What else is it used for?

3 What is meant by 'respiration'?

students' book page 22

B2 2.4 Why do plants need minerals?

KEY POINTS

1 Plants produce sugars through photosynthesis.
2 However, just like animals, they need other substances to grow properly.

Plant roots take up mineral salts for healthy growth.

Nitrates are taken from the soil for producing amino acids. These are used to make proteins for growth. A plant that does not take up enough nitrate (is nitrate deficient) will have stunted growth.

Plants also take up magnesium that is essential to produce chlorophyll. If the plant is deficient in chlorophyll it will have yellow leaves.

Key words: nitrate, magnesium, deficient

The plants on the left of this picture have been grown in a mixture containing all the minerals they need. The experimental plants on the right have been grown without nitrates. The difference in their rate of growth is clear to see.

GET IT RIGHT!

Many students in an exam forget the symptoms of nitrate and magnesium deficiency, or get them the wrong way round. Try to work out a way of memorising them, e.g. nite/stunt and yellow/mag!

CHECK YOURSELF

1 How do plants take minerals from the soil?

2 Why do plants need magnesium?

3 What does 'deficiency' mean?

B2 2 End of chapter questions

1 **How is light energy trapped by a plant?**

2 **You are growing some cuttings in the greenhouse. It is a cold, sunny winter's day. What factor will be limiting the growth of the cuttings?**

3 **Why is glucose stored as an insoluble compound?**

4 **You grow two plants of the same species in the same conditions but one is deficient in nitrate. How will its growth compare with the growth of the healthy plant?**

5 **Which carbohydrate is produced directly as a result of photosynthesis?**

6 **Other than in woodland, where might light be a limiting factor for some plants?**

7 **Why do plants not grow properly if there is not enough nitrate in the soil?**

8 **How do most plants store glucose?**

B2 3 Pre Test: Energy flows

1. **What is a pyramid of biomass?**
2. **Why does a pyramid of biomass often give a more accurate picture of what is happening than a pyramid of numbers?**
3. **Why is energy lost between the different stages of a food chain?**
4. **Why is more energy lost when birds are part of the food chain?**
5. **Why do we say, in terms of food production, that meat is ‘expensive’ to produce?**
6. **How can you limit the energy losses in a food chain?**
7. **What other word do we use for the breakdown of dead plants and animals?**
8. **Which organisms break down dead and waste material?**
9. **Which process results in carbon dioxide being taken out of the atmosphere?**
10. **What do we mean by a ‘stable’ community in terms of the recycling of nutrients?**

students' book page 28

B2 3.1 Pyramids of biomass

KEY POINTS

1. A pyramid of biomass represents the mass of all the organisms at each stage in a food chain.
2. A pyramid of biomass is likely to give a more accurate picture than a pyramid of numbers.

Biomass is the mass of living material in plants and animals.

A pyramid of biomass represents the mass of the organisms at each stage in a food chain. It may be more accurate than a pyramid of numbers. For example, one bush may have many insects feeding on it but the mass of the bush is far greater than the mass of the insects.

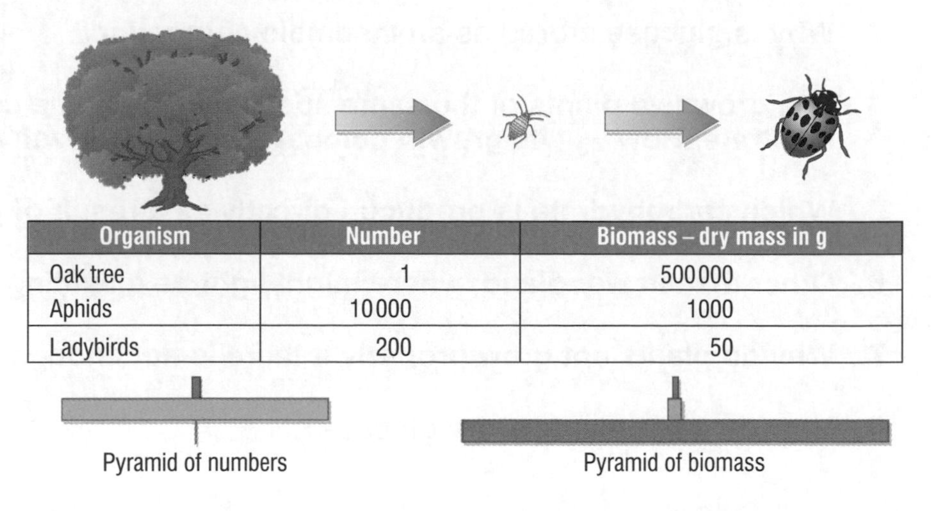

Organism	Number	Biomass – dry mass in g
Oak tree	1	500000
Aphids	10000	1000
Ladybirds	200	50

This food chain cannot be accurately represented using a pyramid of numbers. Using biomass shows us the amount of biological material involved at each level in a way that simple numbers cannot do.

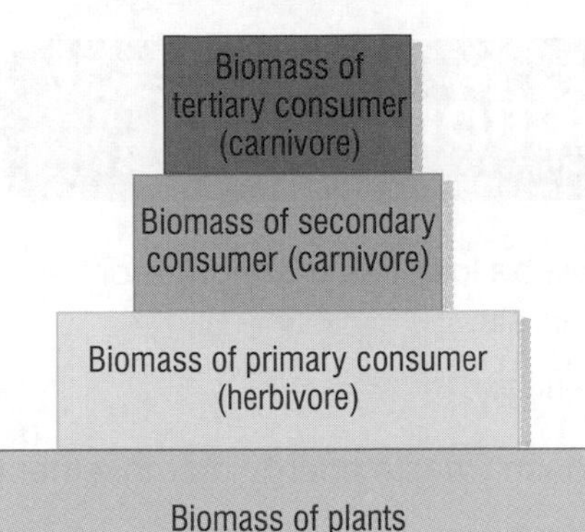

Any food chain can be turned into a pyramid of biomass like this

You can draw pyramids of biomass to scale to give an even more accurate picture.

Key words: pyramid, biomass

Make sure you can explain why a pyramid of biomass is likely to give a more accurate picture of a food chain than a pyramid of numbers.

CHECK YOURSELF

1 What does the word 'biomass' mean?
2 Can you think of another example where a pyramid of biomass will give a more accurate picture of a food chain than a pyramid of numbers?
3 What does drawing a pyramid 'to scale' mean?

students' book page 30

B2 3.2 Energy losses

KEY POINTS

1 For a whole range of reasons, there is energy loss between each stage of a food chain.
2 This means that not all of the energy taken in by an organism results in the growth of that organism.

Not all of the food eaten can be digested, so energy is lost in faeces (waste materials).

Some of the energy is used for respiration, which releases energy for living processes. This includes movement, so the more something moves the more energy it uses and the less is available for growth.

In animals that need to keep a constant temperature, energy from the previous stage of the food chain is used simply to keep the animal at the correct temperature (e.g. 37°C in humans).

AQA EXAMINER SAYS...

Make sure that you can interpret Sankey diagrams. You may well get one in the exam.

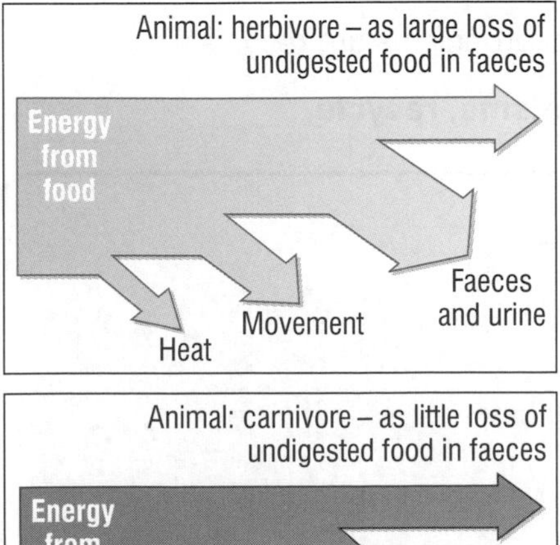

GET IT RIGHT!

Energy is never really 'lost'. What we mean here is that all of the energy in one stage of the food chain does not result in the growth of organisms in the next stage.

Sankey diagrams show how energy is transferred in a system. We can use them to look at the energy which goes in to and out of an animal and predict whether it eats plants or is a carnivore.

CHECK YOURSELF

1 What is meant by 'energy loss' in a food chain?
2 Why does maintaining a constant body temperature use a lot of energy from food?
3 Why does a running horse need more energy than one eating grass in a field?

students' book page 32

B2 3.3 Energy in food production

KEY POINT

If we want to produce food from animals efficiently, then we have to limit the amount of energy they need while they are growing.

AQA EXAMINER SAYS...

There are issues in this topic. One is that if we all ate plants it might mean more food would be available across the world. A second is that if we keep animals so that they produce meat efficiently for us this could be cruel to the animals. Make sure that you have an opinion, but also that you can give both sides of the argument.

The shorter the food chain, the less energy will be lost. It is therefore more efficient for us to eat plants than it is to eat animals.

We can artificially produce meat more efficiently by:

- Preventing the animal from moving so it doesn't waste energy on movement.

This is seen as cruelty by many people and is controversial.

- Keeping the animal at a warmer temperature so it doesn't use as much energy from food to keep itself at a constant temperature.

Key words: efficient

CHECK YOURSELF

1. What do we mean by 'more efficient' food production?
2. If an animal has a constant internal temperature, why might it use more energy when grazing in a field in winter?
3. Why are some methods of producing meat more efficiently said to be cruel?

students' book page 34

B2 3.4 Decay

KEY POINT

All organisms take up nutrients. If they didn't eventually release them the nutrients would run out.

GET IT RIGHT!

If microorganisms are feeding from waste and dead material, then this is how they get their energy. The process of releasing this energy is, therefore, respiration so carbon dioxide is given off.

Detritus feeders (such as worms) may start the process of decay by eating dead animals or plants and producing waste materials. Decay organisms then break down the waste and dead plants and animals.

Decay organisms are microorganisms (bacteria and fungi). Decay is faster if it is warm and wet.

All of the materials from the waste and dead organisms are recycled.

Key words: detritus, decay, microorganisms, recycle

CHECK YOURSELF

1. What is a detritus feeder?
2. Why is decay faster when it is warm?
3. Suggest another word or phrase for 'decay'.
4. Give one example of a microorganism.

students' book page 36

B2 3.5 The carbon cycle

KEY POINT

The recycling of carbon involves both photosynthesis and respiration.

AQA EXAMINER SAYS...

If you can remember that the carbon cycle involves both photosynthesis and respiration, then you will be awarded most of the marks in an exam question.

CHECK YOURSELF

1 Which process returns CO_2 to the atmosphere?

2 Which organisms break down waste products?

3 What does 'recycling carbon' mean?

Photosynthesis removes CO_2 from the atmosphere.

Green plants as well as animals respire. This returns CO_2 to the atmosphere.

Animals eat green plants and build the carbon into their bodies. When plants or animals die (or produce waste) microorganisms release the CO_2 back into the atmosphere through respiration.

A stable community recycles all of the nutrients it takes up.

Key words: photosynthesis, respiration, recycles, stable

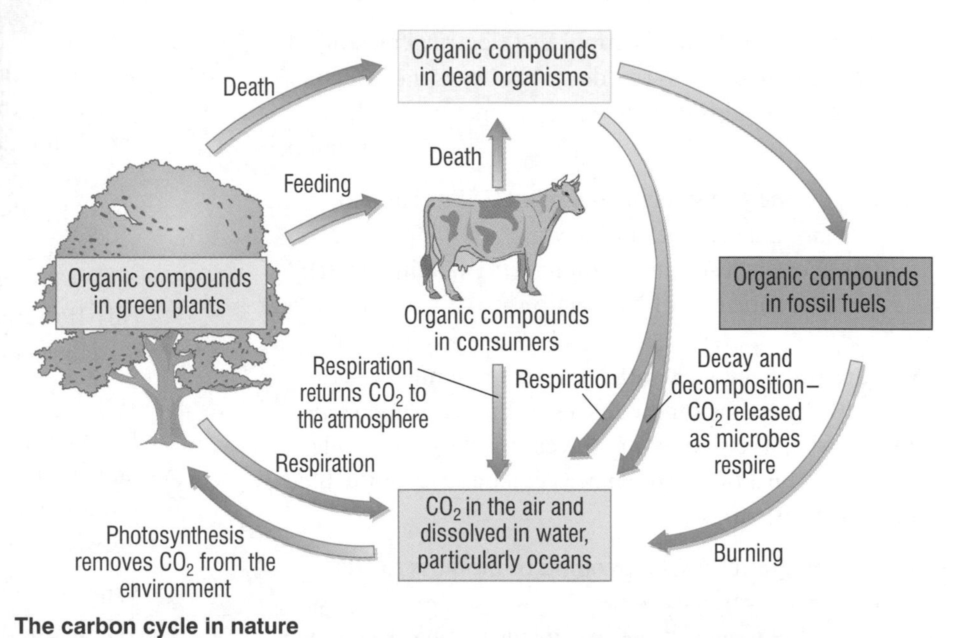

The carbon cycle in nature

B2 3 End of chapter questions

1 **What process is necessary to provide the energy needed to keep humans at a constant temperature?**

2 **Why is very little meat eaten by poorer people in developing countries?**

3 **Why do microorganisms recycle nutrients more quickly when it is warm?**

4 **What involvement do living plants have in the carbon cycle?**

5 **Why does a pyramid of biomass usually give a more accurate picture of what is happening in a community than a pyramid of numbers?**

6 **What happens to the food that animals do not digest?**

7 **Which group of organisms often start the decay process by eating dead plants and animals?**

8 **At what time of the day do plants respire?**

EXAMINATION-STYLE QUESTIONS

1 The diagram shows a nerve cell.

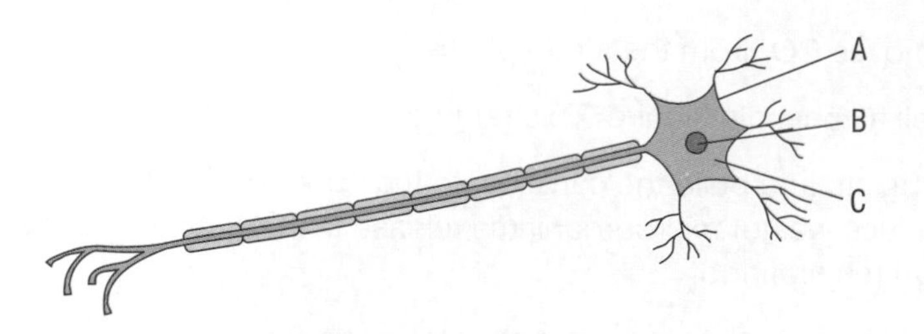

(a) Name parts A, B and C. (3 marks)

(b) The nerve cell carries impulses (electrical messages) around the body. Write down two ways the structure of the cell helps it carry out this function (job). (2 marks)

(c) (i) Name two structures a plant cell has and an animal cell does not have.
(ii) For each structure you have named in part (i), describe one of its functions. (4 marks)

2 You are walking in a wood. You see large numbers of caterpillars feeding on the leaves of the trees and some small birds (blue tits) eating the caterpillars. Suddenly a fairly large bird (a sparrowhawk) catches one of the blue tits and starts eating it.

(a) The diagram shows a pyramid of biomass for the organisms you have seen.
Write the names of the organisms you have seen in the woods in the correct places next to the pyramid of biomass. (3 marks)

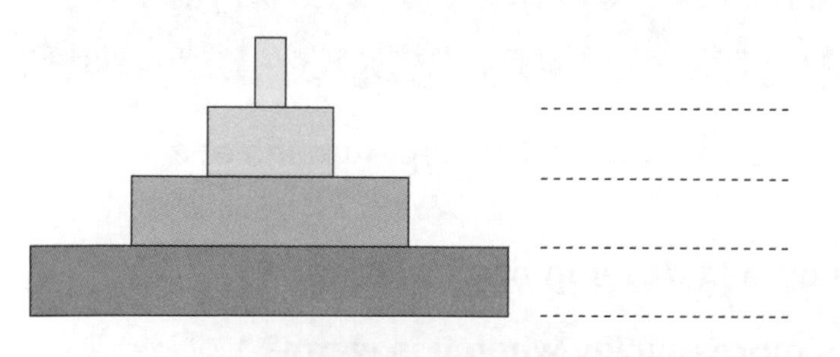

(b) Why is so much energy lost between each of the stages in this woodland food chain? (3 marks)

(c) Producing meat more efficiently means that more of the food that the animals eat is converted into meat and not wasted. How can we produce meat more efficiently? (2 marks)

3 The photograph shows a very small area of the Amazon jungle in Ecuador, South America. The growth of plants is very fast in the jungle.

(a) The plants in the jungle photosynthesise and produce glucose in their leaves.
(i) What do the plants use the glucose for? (1 mark)
When does this process take place? (1 mark)
(ii) The plants will also be taking up nitrates from the soil.
What do they use these nitrates for? (2 marks)
(iii) If there is not enough nitrate then the plant will have a deficiency. What will the plant look like? (1 mark)

(b) In an area where plants are growing so quickly, they may not be able to take up enough magnesium.
Explain the results of magnesium deficiency in a growing plant. (2 marks)

(c) When the plants die they are recycled.
What conditions increase the speed at which recycling takes place? (2 marks)

(d) Except where humans are present, the Amazon jungle is said to be a 'stable' community. What does this mean? (1 mark)

4 Cells need to take up materials that they need and get rid of waste materials from the reactions that have taken place.

(a) (i) Diffusion is one way that materials pass into and out of cells.
What is meant by diffusion? (3 marks)
(ii) In a plant cell which is photosynthesising which gas, overall, will be:
A diffusing into the plant cells
B diffusing out of the plant cells? (2 marks)

(b) Why do the cells not need to provide energy for the process of diffusion? (2 marks)

(c) Osmosis is a 'special' case of diffusion.
(i) How is osmosis similar to diffusion? (2 marks)
(ii) How is osmosis different to diffusion? (2 marks)

The answer is worth 3 marks out of a possible 4. The responses worth a mark are underlined in red.

We can improve the answer in several ways:

It would improve the answer to state that the light provides the energy for the process.

The comment is too vague to be worth a mark. A fuller explanation is required, e.g. the greenhouse will be warm as it is sunny, the enzymes controlling photosynthesis will, therefore, be working quickly.

The diagram shows a greenhouse on a sunny summer's day.

Which one of these factors will limit the rate of photosynthesis in the greenhouse?

- **the level of carbon dioxide**
- **the level of light**
- **the temperature**

Explain your answer. *(4 marks)*

The carbon dioxide level.
The plants will photosynthesise quickly as there should be enough light as it is a sunny day. The temperature will also be warm enough. If all of these things are going quickly enough then the plants will use up the carbon dioxide quickly so there won't be enough.

The answer is worth 2 marks out of the 4 marks available. The responses worth a mark are underlined in red.

We can improve the answer in several ways:

In the Autumn many plants lose their leaves for the winter. There are carbon compounds in the leaves. How is this carbon recycled so it can be used again by plants? *(4 marks)*

When the leaves fall little animals come and eat them and through respiration energy and carbon dioxide are released. Then other little animals feed on the waste and the whole process is repeated.

The 'little animals' mentioned would gain the mark if 'bacteria or fungi', or the general term 'microorganisms', had been given.

'Detritus feeders' feed on the leaves (digest them) and start the process of breakdown, e.g. earthworms.

The process of breakdown is also known as 'decomposition' or 'decay'.

B2 | Additional biology (Chapters 4–6)

Checklist

This spider diagram shows the topics in the second half of the unit. You can copy it out and add your notes and questions around it, or cross off each section when you feel confident you know it for your exams.

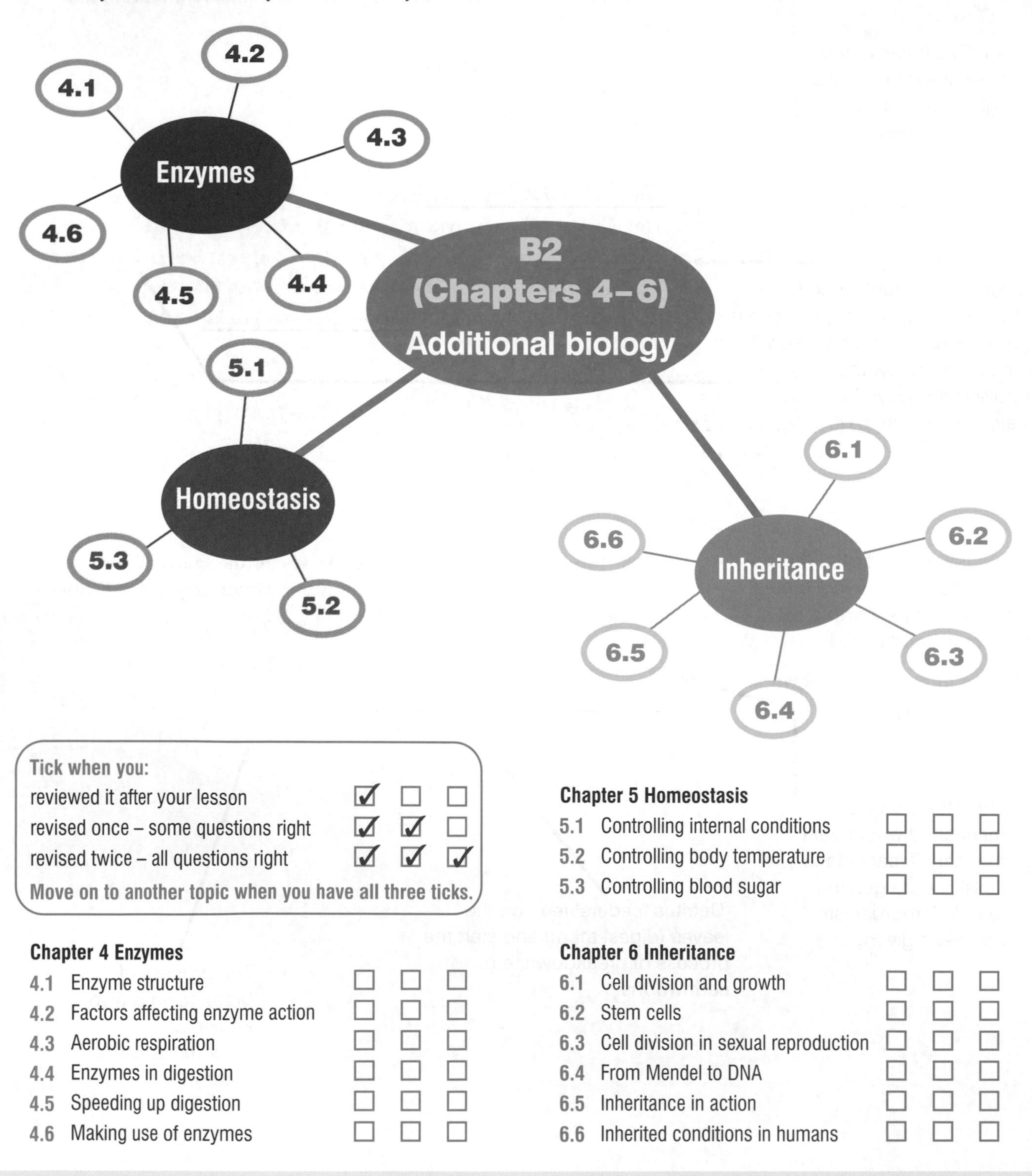

Tick when you:

reviewed it after your lesson	☑	☐	☐
revised once – some questions right	☑	☑	☐
revised twice – all questions right	☑	☑	☑

Move on to another topic when you have all three ticks.

Chapter 4 Enzymes

4.1	Enzyme structure	☐	☐	☐
4.2	Factors affecting enzyme action	☐	☐	☐
4.3	Aerobic respiration	☐	☐	☐
4.4	Enzymes in digestion	☐	☐	☐
4.5	Speeding up digestion	☐	☐	☐
4.6	Making use of enzymes	☐	☐	☐

Chapter 5 Homeostasis

5.1	Controlling internal conditions	☐	☐	☐
5.2	Controlling body temperature	☐	☐	☐
5.3	Controlling blood sugar	☐	☐	☐

Chapter 6 Inheritance

6.1	Cell division and growth	☐	☐	☐
6.2	Stem cells	☐	☐	☐
6.3	Cell division in sexual reproduction	☐	☐	☐
6.4	From Mendel to DNA	☐	☐	☐
6.5	Inheritance in action	☐	☐	☐
6.6	Inherited conditions in humans	☐	☐	☐

What are you expected to know?

Chapter 4 Enzymes

See students' book pages 42–57

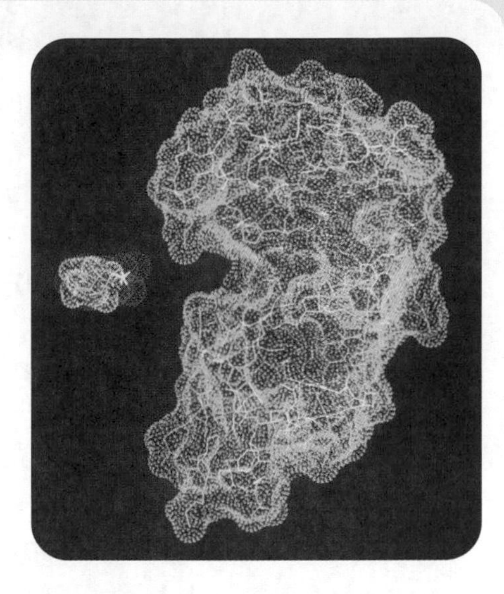

- How enzymes work in terms of their structure.
- That different enzymes work best under different conditions of pH.
- How aerobic respiration releases energy.
- The role of enzymes in the digestion of carbohydrate, protein and fat (lipids).
- The role of the stomach, pancreas, small intestine and liver in the digestion of food.
- How enzymes are used in the home.
- How enzymes are used in industry.

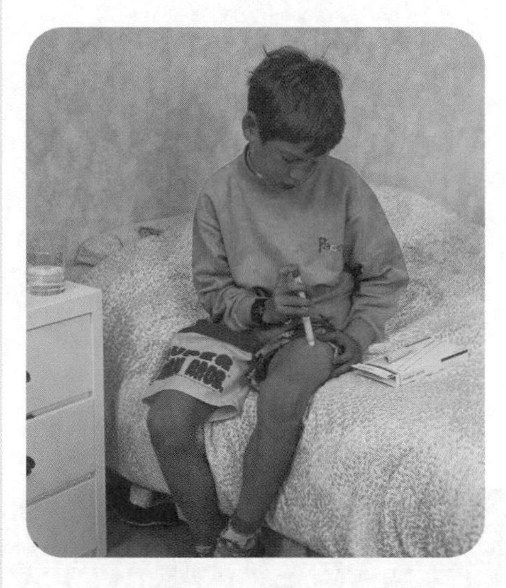

Chapter 5 Homeostasis

See students' book pages 58–67

- How we get rid of the waste products carbon dioxide and urea.
- How we keep our internal conditions constant including:
  - water content
  - ion content
  - temperature
  - blood sugar level.

Chapter 6 Inheritance

See students' book pages 68–83

- How characteristics are passed on from one generation to the next (inheritance).
- What happens in mitosis.
- What happens during meiosis. [Detail and comparisons with mitosis are Higher Tier only]
- The issues surrounding the use of stem cells (from embryos and adult bone marrow).
- That a gene is a short length of DNA controlling one characteristic, that pairs of genes controlling the same characteristic are known as 'alleles' and that chromosomes are made up of many genes.
- That some disorders, e.g. Huntingdon's disease and cystic fibrosis, are inherited.
- That embryos can be screened for these disorders and other genetic disorders.
- How to interpret genetic diagrams.
- How to draw genetic diagrams. [Higher Tier only]

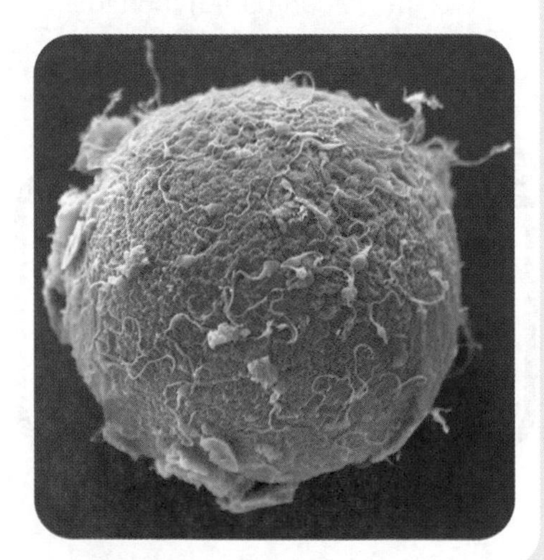

B2 4 Pre Test: Enzymes

1. What do we mean by the term 'biological catalyst'?
2. State two processes that are controlled by enzymes.
3. Why is the shape of an enzyme important?
4. Where does aerobic respiration take place?
5. In living cells what are amino acids built up into?
6. What is the general name for an organ which produces an enzyme?
7. What type of enzyme breaks down fats?
8. Where is bile produced?
9. Where is hydrochloric acid produced?
10. Why is fructose such a useful sugar?

students' book page 42

B2 4.1 Enzyme structure

KEY POINTS

1. The structure of an enzyme allows certain molecules to fit.
2. If this structure is changed (the enzyme is denatured) then the enzyme cannot work as a catalyst.

GET IT RIGHT!

Many students in the exam talk of enzymes being 'killed' by high temperatures. This answer is worth *no marks* – they are destroyed or denatured.

Enzymes are biological catalysts – they speed up reactions.

Enzymes are large proteins and each has a particular shape. This shape has an area where other molecules can fit in. This area is called the 'active site'.

Too high a temperature will change the enzymes shape, and it will no longer work. We say it has been destroyed or denatured.

Enzymes can catalyse the build up of small molecules into large molecules or the break down of large molecules into small molecules.

Enzymes lower the amount of energy necessary for a reaction to take place – the 'activation' energy.

Key words: catalyst, active site, denatured, activation energy

CHECK YOURSELF

1. What type of molecules are enzymes?
2. What is the 'active site'?
3. What does 'catalysis' mean?

students' book page 44

B2 4.2 Factors affecting enzyme action

KEY POINT

Enzymes are very sensitive to both temperature and pH.

AQA EXAMINER SAYS...

If you get a question about rate of reaction, remember to talk about collisions. How hard and how often molecules collide determines the rate of reaction. Don't be vague – mention 'collisions'!

Reactions take place faster when it is warmer. At the higher temperature the molecules move around more quickly so collide with each other more often and with more energy.

However, if the temperature gets too hot the enzyme stops working. That's because the active site changes shape and the enzyme becomes denatured.

Enzymes work best in certain acidic or alkaline conditions (pH). If the pH is too acidic or alkaline for the enzyme, then the active site could change shape. The enzyme would stop working.

Key words: collide, active site

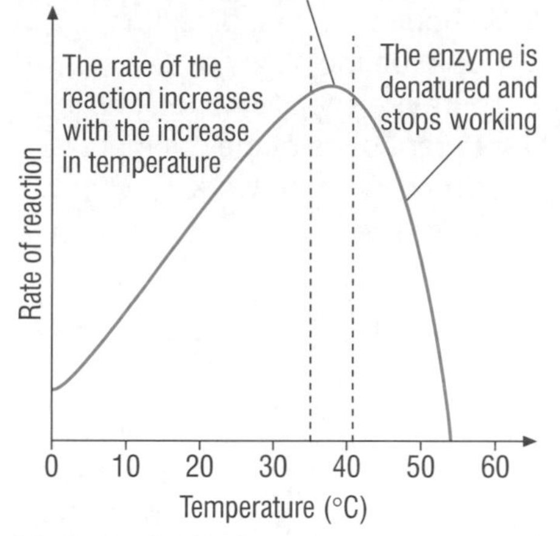

Like most chemical reactions, the rate of an enzyme controlled reaction increases as the temperature rises – but only until the point where the complex protein structure of the enzyme breaks down

CHECK YOURSELF

1 Why does an increase in temperature increase the rate of reaction?

2 Why do enzymes stop working if the temperature is too hot?

3 Why does the pH have to be 'right' for a reaction to go well?

students' book page 46

B2 4.3 Aerobic respiration

KEY POINT

Aerobic respiration is the release of energy from food when oxygen is available.

EXAM HINT

Remember that animals and plants both respire, it is the same process. Plants do not just photosynthesise.

The equation for respiration is:

glucose + oxygen → carbon dioxide + water **[+energy]**

The process mostly takes place in the mitochondria.

The energy released is used to:

- build larger molecules from smaller ones
- enable muscle contraction in animals
- maintain a constant body temperature in mammals and birds
- build sugars, nitrates and other nutrients in plants into amino acids and then proteins.

Key words: mitochondria, energy

CHECK YOURSELF

1 Where does aerobic respiration take place?

2 What is needed for muscles to contract?

3 What are amino acids built up into?

students' book page 48

B2 4.4 Enzymes in digestion

KEY POINTS

1 Without enzymes, digestion would be too slow.
2 There are specific conditions in different parts of the gut that help enzymes to work effectively.

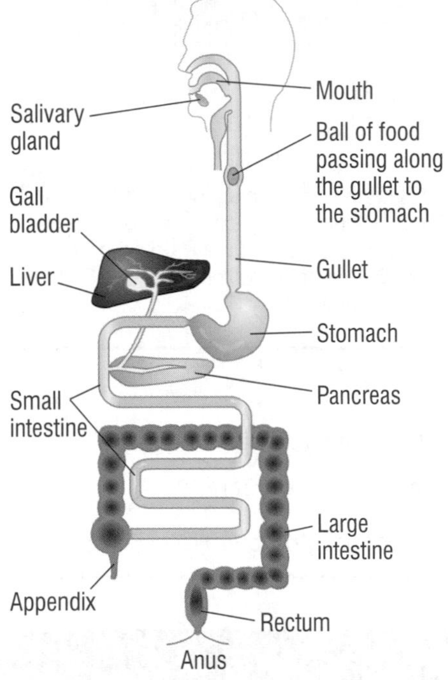

The human digestive system

Digestion involves the breakdown of large, insoluble molecules into smaller soluble molecules.

- Amylase (a carbohydrase) is produced by the salivary glands, the pancreas and the small intestine. Amylase catalyses the digestion of starch into sugars in the mouth and small intestine.
- Protease is produced by the stomach, the pancreas and the small intestine. Protease catalyses the breakdown of proteins into amino acids in the stomach and small intestine.
- Lipase is produced by the pancreas and small intestine. Lipase catalyses the breakdown of lipids (fats and oils) to fatty acids and glycerol.

Key words: amylase, protease, lipase, lipid, amino acid, fatty acid, glycerol

Try to remember which enzyme does what, where it is produced and where it works. There are a lot of marks to lose in these questions if you don't remember the facts.

CHECK YOURSELF

1 What are lipids?

2 Which enzymes does the pancreas produce?

3 What are the products of lipase digestion?

students' book page 50

B2 4.5 Speeding up digestion

KEY POINT

Enzymes only work well within a narrow pH range.

GET IT RIGHT!

Large numbers of students do not remember that bile is made by the liver and stored in the gall bladder. Try to get it right!

- Protease enzymes in the stomach work best in acid conditions. Glands in the stomach wall produce hydrochloric acid to create very acidic conditions.
- Amylase and lipase work in the small intestine. They work best when the conditions are slightly alkaline.
- The liver produces bile that is stored in the gall bladder. Bile is squirted into the small intestine and neutralises the stomach acid. It makes the conditions slightly alkaline.

Key words: acid, alkali, bile, liver, gall bladder

CHECK YOURSELF

1 What conditions are best in the small intestine?

2 What is the function of bile?

3 What type of acid is produced in the stomach?

students' book page 52

B2 4.6 Making use of enzymes

KEY POINTS

1. Enzymes can be used in products in the home and in industry.
2. Microorganisms produce enzymes that we can use.

Biological washing powders contain enzymes that digest food stains. They work at lower temperatures than ordinary washing powders so can save us money.

We also use:

- Protease enzymes to pre-digest proteins in some baby foods.
- Isomerases to convert glucose into fructose. Fructose is much sweeter, so less is needed in foods. The foods, therefore, are not so fattening.
- Carbohydrases to convert starch into sugar syrup for use in foods.

Key words: isomerase, fructose, pre-digest

Learning to eat solid food isn't easy. Having some of it pre-digested by protease enzymes can make it easier to get the goodness you need to grow!

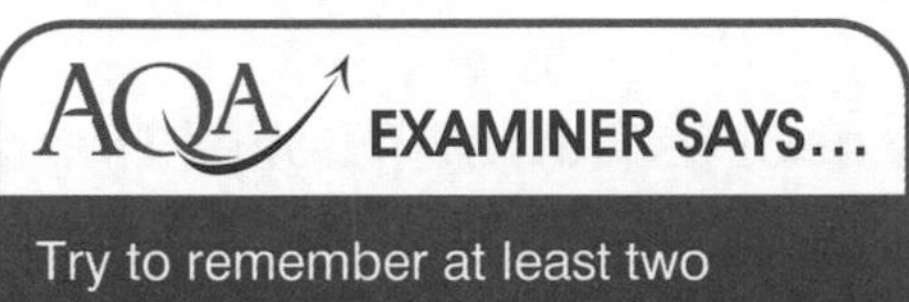

Try to remember at least two examples of these uses of enzymes. Most questions will not ask for more than two.

CHECK YOURSELF

1. Why is fructose used in slimming foods?
2. Suggest why baby food might need some of the protein content 'pre-digested'.
3. Why would you not use a temperature of above about 45°C if you are using a biological washing powder?

B2 4 End of chapter questions

1. **What do we mean by the 'activation energy'?**
2. **What is the 'active site' of an enzyme?**
3. **State two uses, in animals, of the energy released by respiration.**
4. **What digestive process takes place in the stomach?**
5. **What term do we use when the active site of an enzyme changes shape?**
6. **What are the two waste products of aerobic respiration?**
7. **Where is protease produced in the digestive system?**
8. **Where is bile made and where is it stored?**

B2 5 Pre Test: Homeostasis

1. What are the waste products of respiration?
2. What happens to the amino acids we cannot use?
3. Where is urea produced?
4. Why is the ion content of the body important?
5. Which two areas of the body detect changes in temperature?
6. Why does sweating more cool us down? [Higher Tier only]
7. Where is sweat produced? [Higher Tier only]
8. What effect does 'shivering' have on the body? [Higher Tier only]
9. Which hormone helps to control the blood sugar level?
10. Where is this hormone produced?

students' book page 58

B2 5.1 Controlling internal conditions

KEY POINTS

1. We must remove the waste products produced through chemical reactions from the body.
2. There are other factors we must keep within certain limits, e.g. water and ion content of the cells.

The processes in your body that help to maintain a constant internal environment are known as homeostasis.

Just think of the different temperatures we experience:

Your body also has to cope with varying rates of respiration needed in different activities:

Whatever you choose to do in life, the conditions inside your body will stay more-or-less exactly the same

We can investigate the effect of exercise on our breathing using the equipment below:

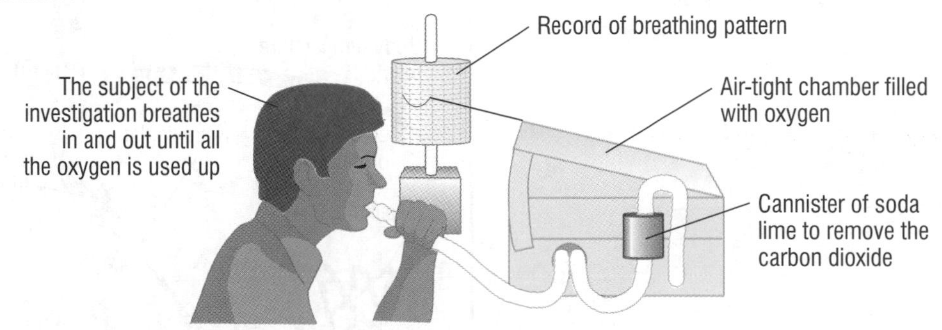

Because you breathe in and out of the machine all the time, you can't get rid of your waste carbon dioxide in the normal way. There has to be a special filter to remove the carbon dioxide so it doesn't poison you!

- Carbon dioxide is a waste product of respiration, it is excreted through the lungs.
- Some of the amino acids we take in are not used. They are converted into urea by the liver and excreted by the kidneys in the urine. Urine can be stored in the bladder.
- The water and ion content of cells must be carefully controlled. If they are not, then too much or too little water may move in and out of cells by osmosis.

Key words: homeostasis, amino acids, urea, urine, liver, kidney, ions, osmosis

GET IT RIGHT!

Urea is produced by the liver and excreted by the kidneys. A lot of exam candidates get this the wrong way round (or forget it altogether!).

CHECK YOURSELF

1 Which process results in the production of carbon dioxide?

2 What are unwanted amino acids converted into?

3 Where is urine stored?

students' book page 60

B2 5.2 Controlling body temperature

KEY POINTS

1 Enzymes work in a very narrow temperature range.
2 We must keep our body temperature within that range.

GET IT RIGHT!

Most students cannot explain how sweating cools you down. Remember that the evaporation of the sweat *takes energy from the skin* – that's why you cool down!

The thermoregulatory centre of the brain and receptors in the skin detect changes in temperature. The thermoregulatory centre controls the body's response to a change in internal temperature.

If the core temperature *rises*:

- Blood vessels near the surface of the skin dilate allowing more blood to flow through the skin capillaries. Heat is lost by radiation.
- Sweat glands produce more sweat. This evaporates from the skin's surface. The energy required for it to evaporate comes from the skin's surface. So we cool down.

If the core temperature *falls*:

- Blood vessels near the surface of the skin constrict and less blood flows through the skin capillaries. Less heat is radiated.
- We 'shiver'. Muscles contract quickly. This requires respiration and some of the energy produced is released as heat.

Key words: thermoregulatory, dilate, constrict, radiation, capillaries

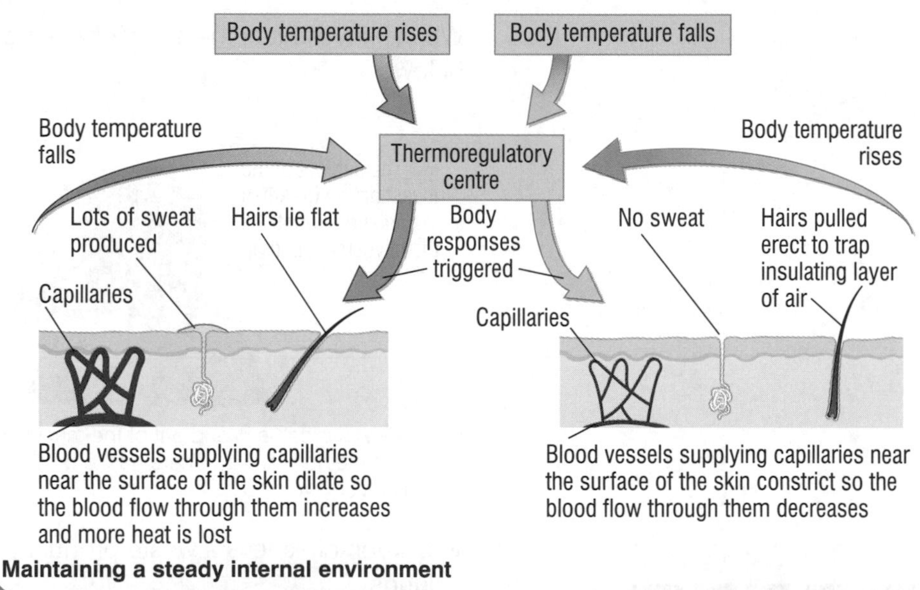

Maintaining a steady internal environment

CHECK YOURSELF

1 Which two parts of the body detect changes in temperature?
2 Where is sweat produced? [Higher Tier only]
3 Why does 'shivering' make you feel warmer? [Higher Tier only]

students' book page 62

B2 5.3 Controlling blood sugar

KEY POINTS

1 The level of sugar in the blood must be kept at the correct level.
2 Hormones help our bodies to do this.

The pancreas monitors and controls the level of sugar in our blood.

If there is too much sugar in our blood the pancreas produces the hormone insulin that results in the excess sugar being stored in the liver as glycogen. If insulin is not produced the blood sugar level may become fatally high.

If the pancreas is not producing enough insulin, this is known as diabetes. It can sometimes be controlled by diet or the person may need insulin injections.

Key words: pancreas, monitor, insulin, glycogen, diabetes

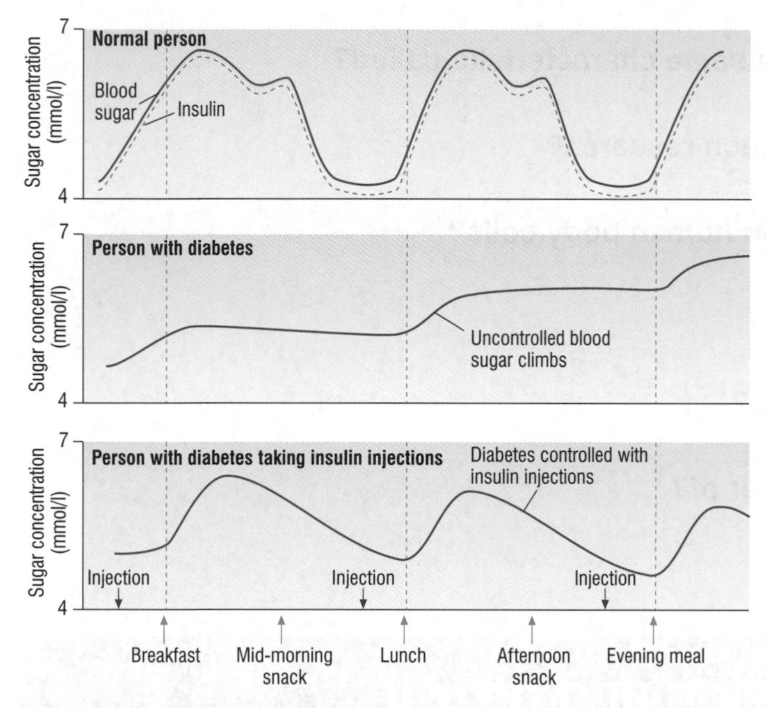

These graphs show the impact insulin injections have on people affected by diabetes. The injections keep the blood sugar level within safe limits.

AQA EXAMINER SAYS…

Remember it is when the blood sugar level is too high that insulin is produced.

CHECK YOURSELF

1 Which organ monitors blood sugar level?
2 Where is glycogen stored?
3 When the pancreas does not produce enough insulin, what is the condition known as?

B2 5 End of chapter questions

1 **Why is the ion content of the cells important?**
2 **What are the roles of the thermoregulatory centre?**
3 **Why does sweating cool you down? [Higher Tier only]**
4 **What happens if, after a large meal, there is too much sugar in the blood?**
5 **When your skin flushes, how is heat lost from the skin? [Higher Tier only]**
6 **How is urea eventually excreted?**
7 **What happens to the blood vessels near the surface of the skin if it is cold? [Higher Tier only]**
8 **If there is an increase in the concentration of substances in a living cell, which process will be affected?**

B2 6 Pre Test: Inheritance

1. **In which cells are chromosomes normally found in pairs?**
2. **When a cell divides by mitosis, how many new cells are formed?**
3. **Why are cells produced by mitosis?**
4. **How many cells are produced from a parent cell in meiosis? [Higher Tier only]**
5. **What are pairs of genes controlling the same characteristic called?**
6. **Why are stem cells the subject of so much research?**
7. **How many pairs of chromosomes are in human body cells?**
8. **What does DNA stand for?**
9. **What do we mean by a 'dominant' gene?**
10. **What is Huntingdon's disease a disorder of?**

students' book page 68

B2 6.1 Cell division and growth

KEY POINTS

1 Body cells need to divide to produce new cells for growth or repair.

2 Mitosis is the type of cell division that produces identical new cells.

Cell division is necessary for the growth of an organism, or for repair if tissues are damaged.

Mitosis results in two identical cells being produced from the original cell.

A copy of each chromosome is made before the cell divides and one of each chromosome goes to each new cell.

Key words: mitosis, growth, repair

GET IT RIGHT!

Mitosis results in two new cells each identical to the parent cell. The new cells are either for growth or replacement (repair).

CHECK YOURSELF

1 Why do cells divide by mitosis?

2 Why is each new cell identical to the parent cell?

3 How many cells are produced in each division by mitosis?

students' book page 70

B2 6.2 Stem cells

KEY POINTS

1 Stem cells are not specialised, but can differentiate into many different types of cell when required.
2 There are ethical issues surrounding the use of stem cells.

BUMP UP YOUR GRADE

There is a lot of argument about the use of embryonic stem cells for research. There are major ethical issues. You could be asked to offer your view and you may have a strong opinion. In the exam, give both sides of the argument if you want to gain full marks.

Stem cells are unspecialised. They can develop (differentiate) into many different types of specialised cell. Stem cells are found in the embryo and in adult bone marrow.

Many embryonic stem cells that we carry research out on are from aborted embryos, or are 'spare' embryos from fertility treatment. This results in ethical issues and much debate, as it can be argued that you are destroying life to obtain these stem cells for research.

The use of stem cells from adult bone marrow is still limited by the number of different types of specialised cell we can develop them into.

Key words: stem cell, specialised, embryonic, bone marrow

CHECK YOURSELF

1 Why are stem cells important?
2 Why are there ethical issues surrounding their use?
3 Why can we not just use stem cells from adult bone marrow?

students' book page 72

B2 6.3 Cell division in sexual reproduction

KEY POINTS

1 Sex cells are produced by meiosis. [Higher Tier only]
2 Four cells are produced from each parent cell. They are all different. [Higher Tier only]

AQA EXAMINER SAYS...

Make sure that you can spell mitosis and meiosis. You may answer a question very well and lose nearly all of the marks, if the examiner cannot tell whether you are talking about mitosis or meiosis.

HIGHER

Cells in reproductive organs, e.g. testes and ovaries, divide to form sex cells (gametes).

Before division, a copy of each chromosome is made. The cell now divides twice to form four gametes (sex cells). This type of cell division is called meiosis.

Each gamete has only one chromosome from the original pair. All of the cells are different from each other and the parent cell.

Sexual reproduction results in variation as the sex cells (gametes) from each parent fuse. So half the genetic information comes from the father and half from the mother.

Key words: meiosis, gametes, variation

CHECK YOURSELF

1 How many sex cells are produced in one meiotic division? [Higher Tier only]
2 How many times does the parent cell divide to produce the gametes? [Higher Tier only]
3 Why does sexual reproduction result in variation?

students' book page 74

B2 6.4 From Mendel to DNA

KEY POINTS

1 Gregor Mendel worked out how characteristics are inherited.
2 Genes make up the chromosomes, which control our characteristics.

Gregor Mendel was a monk who worked out how characteristics were inherited. His ideas were not accepted for many years.

Genes are short lengths of DNA (deoxyribonucleic acid), which make up chromosomes and control our characteristics.

Genes code for combinations of specific amino acids, which make up proteins.

Key words: genes, DNA, chromosomes

CHECK YOURSELF

1 What was the name of the monk who worked out the patterns of inheritance?
2 What does DNA stand for?
3 What are genes made up of?

students' book page 76

B2 6.5 Inheritance in action

KEY POINTS

1 Alleles control the development of characteristics.
2 Some alleles are dominant and some are recessive.

- Human beings have 23 pairs of chromosomes, one pair are the sex chromosomes. Females are XX and males XY.
- Genes controlling the same characteristic are called alleles.
- If an allele 'masks' the effect of another it is said to be 'dominant'. The allele where the effect is 'masked' is said to be 'recessive'.

For example, the allele for brown eyes is dominant to the allele for blue eyes.

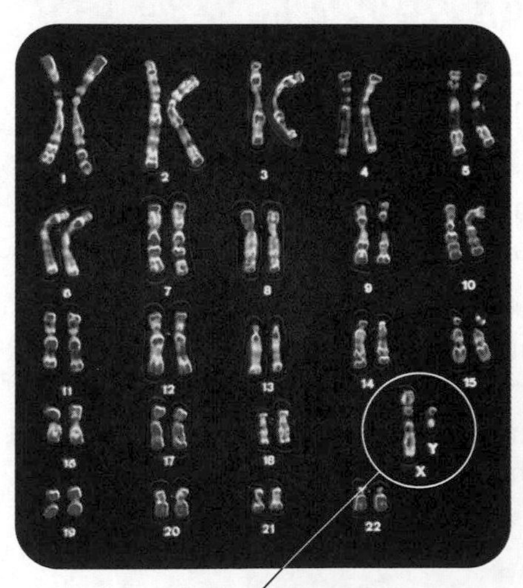

The chromosomes of the human male

HIGHER

If two parents have brown eyes and have the genetic make up Bb, what would be the chance of them having a blue eyed child?

Parents	Bb	Bb
Sex cells	B or b	B or b

Fertilisation →

	B	b
B	BB	Bb
b	bB	**bb**

There is a 1 in 4 (25%) chance of having a blue eyed child (bb)

Key words: allele, dominant, recessive

You can draw Punnett Squares to show genetic crosses or 'line' diagrams. It is easier to make mistakes with 'line' diagrams and also harder for the examiner to see what you are doing.

CHECK YOURSELF

1 How many chromosomes does a human being have?
2 What are the sex chromosomes of a male?
3 What is meant by a 'recessive allele'?

students' book page 78

B2 6.6 Inherited conditions in humans

KEY POINTS

1. Huntington's disease is caused by a dominant allele.
2. Cystic fibrosis is caused by a recessive allele.
3. Embryos can be screened for these conditions and other genetic disorders.

Huntington's disease is a disorder of the nervous system. It is caused by a dominant allele, so even if only one parent has the disease it can be inherited by a child.

Cystic fibrosis is a disorder of cell membranes. It is caused by a recessive allele so parents may be carriers (Cc). Only if both parents are either carriers or have the disorder does a child inherit it.

Embryos can be screened to see if they carry alleles for one of these or other genetic disorders.

Key words: disorder, Huntington's, cystic fibrosis, carrier

If you are taking the Higher Tier exam, try to practise doing genetic diagrams to show how each disorder can be inherited. Remember a Punnett Square is the easiest way to show your results.

HIGHER

Both parents are carriers, so Cc

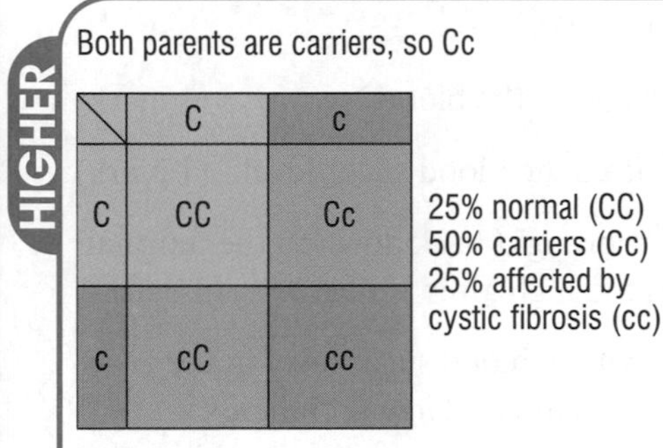

	C	c
C	CC	Cc
c	cC	cc

3/4, or 75% chance normal
1/4, or 25% chance cystic fibrosis

The faulty alleles can be covered up by normal alleles for generations until two carriers have a child and, by chance, both of the cystic fibrosis alleles are passed on

Parent with Huntington's disease Hh
Normal parent hh

	H	h
h	Hh	hh
h	hH	hh

50% chance Huntington's disease, Hh or hH
50% chance normal, hh

A genetic diagram for Huntington's disease shows how a dominant allele can affect offspring

CHECK YOURSELF

1. What system in the body does Huntington's disease affect?
2. What is meant by a 'carrier'?
3. When embryos are screened, what are the scientists looking for?

B2 6 End of chapter questions

1. **When we say that stem cells 'differentiate', what do we mean?**
2. **What is the result of one meiotic cell division? [Higher Tier only]**
3. **What is meant by 'dominant' when we refer to alleles?**
4. **What is a 'gene'?**
5. **What do the chromosomes do before cell division takes place?**
6. **How many pairs of chromosomes are present in human cells (other than sex cells)?**
7. **What are 'alleles'?**
8. **What is cystic fibrosis a disorder of?**

EXAMINATION-STYLE QUESTIONS

1 We need to keep the internal conditions in our bodies constant. The diagram shows some of the organs that do this.

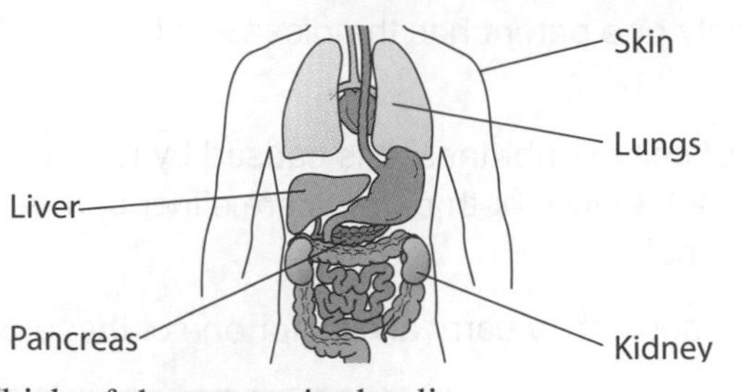

(a) Which of the organs in the diagram:
 (i) excretes carbon dioxide
 (ii) produces urea
 (iii) excretes urea in the urine
 (iv) produces sweat? (4 marks)

(b) How does sweat help to cool the body down on a hot day? (3 marks)

(c) On a cool day we might 'shiver'. How does 'shivering' help us to become warmer? [Higher] (2 marks)

2 (a) Where are stem cells found? (2 marks)

(b) Why is research into stem cells so important? (2 marks)

(c) Why is there so much argument about whether research into stem cell research should be taking place at all? (3 marks)

3 Meat is protein. Proteins are digested by protease enzymes.

A student investigated the effect of pH on the digestion of cubes of meat by protease enzymes. The student kept all of the other variables the same. The table shows the student's results.

pH	Time taken to digest the meat (minutes)
1	14
2	12
3	18
4	32
5	52
6	lesson had ended

(a) (i) What pH provided the best conditions for the enzyme to work in? (1 mark)
 (ii) From the information given above, how could the student have improved the investigation? (1 mark)
 (iii) What further work could the student carry out to find out more accurately the best pH for this enzyme? (2 marks)

(b) In which part of the gut do you think this protease enzyme is found? Explain your answer. (2 marks)

(c) (i) What do the following enzymes digest and what are the products of this digestion?
 • lipase • amylase (5 marks)
 (ii) Where is the enzyme amylase produced? (2 marks)

(d) What is the function (job) of bile? (2 marks)

4 It is very important that we keep our blood sugar at a fairly constant level in the body.

(a) Why do we need sugar in the blood? (1 mark)

(b) Which organ monitors the blood sugar level? (1 mark)

(c) How is blood sugar brought back down to the 'normal' level if it becomes too high after a meal? (3 marks)

(d) If you cannot control the blood sugar level in the body you may suffer from diabetes. How is diabetes controlled? (2 marks)

5 Huntington's disease is a rare disorder of the nervous system. It is inherited through a dominant allele represented by H. The recessive allele of this pair of genes is represented by h.

Study the diagram below, which shows the inheritance of Huntington's disease in a family.

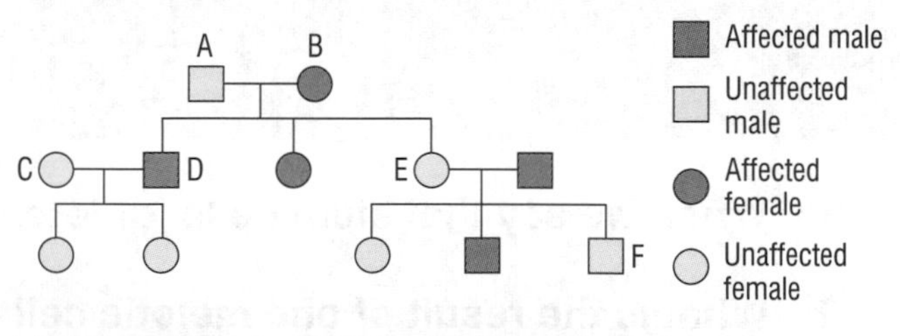

(a) Use a genetic diagram to show why D has Huntington's disease. (3 marks)

(b) Why does E not have the disease? (2 marks)

(c) If F were to have children with a person carrying one allele for Huntington's disease, what would be the chance of a child having the disease? Explain your answer. (2 marks)

(d) (i) How does DNA control the inheritance of characteristics of organisms? (2 marks)
 (ii) What is a short length of DNA coding for one characteristic called? [Higher] (1 mark)

EXAMPLES OF EXAM ANSWERS

This is a very difficult question (Higher Tier only) and the student has gained 3 of the 4 marks available.

The responses worth a mark are underlined in red.

We can improve the answer in several ways:

The sex cells are produced by meiosis.
Describe what happens to the chromosomes during meiosis and how meiosis and sexual reproduction lead to variation in the offspring. *(4 marks)*

During meiosis each chromosome splits itself in half and copies each half. They then all line up in the middle of the cell and are separated by strands from the cell. In the new cells they double again and repeat the process.

During fertilisation we do not know which of the sex cells (copies) will fuse together. Each instruction is taken from the male and female chromosome, even then the alleles could be dominant or recessive.

In meiosis the chromosomes do copy, but don't 'split in half'.

Then one from each pair goes to each of the two new cells.

Many sex cells are produced and which sperm fertilises which egg is a random process. The sex cells will carry the genes from each parent and the candidate is correct in that dominant or recessive alleles may be carried.

The answer is worth 4 marks out of the 6 marks available.

The responses worth a mark are underlined in red.

The student has used the information well and gains all 4 of the marks for the first section. Most students should be able to use information presented in this way.

In the second part the student scores no marks.

We can improve the answer in several ways:

We use enzymes in industry. Here are some properties of enzymes.

- **They work at low temperatures so can save energy and, therefore, money.**
- **Some reactions require a high pressure, enzymes lower the pressure necessary.**
- **They are easily broken down at too high a temperature or the wrong pH.**
- **They are soluble in water, so are difficult to separate when a reaction has finished.**
- **They are produced from microorganisms, so are expensive to buy.**

(a) Give two advantages and two disadvantages of using enzymes in industry? *(4 marks)*

(b) Why don't enzymes work at high temperatures? *(2 marks)*

(a) Advantages are they work at low temperatures so save money and low pressure so you don't need such expensive equipment. Disadvantages are that they cost a lot and it is difficult to get them back after the reaction.

(b) They don't work at high temperatures as they are killed.

The enzymes don't work as their shape is altered. They are therefore denatured (or destroyed). Never say that enzymes are killed – they are chemicals!

C2 | Additional chemistry (Chapters 1–3)

Checklist

This spider diagram shows the topics in the first half of the unit. You can copy it out and add your notes and questions around it, or cross off each section when you feel confident you know it for your exams.

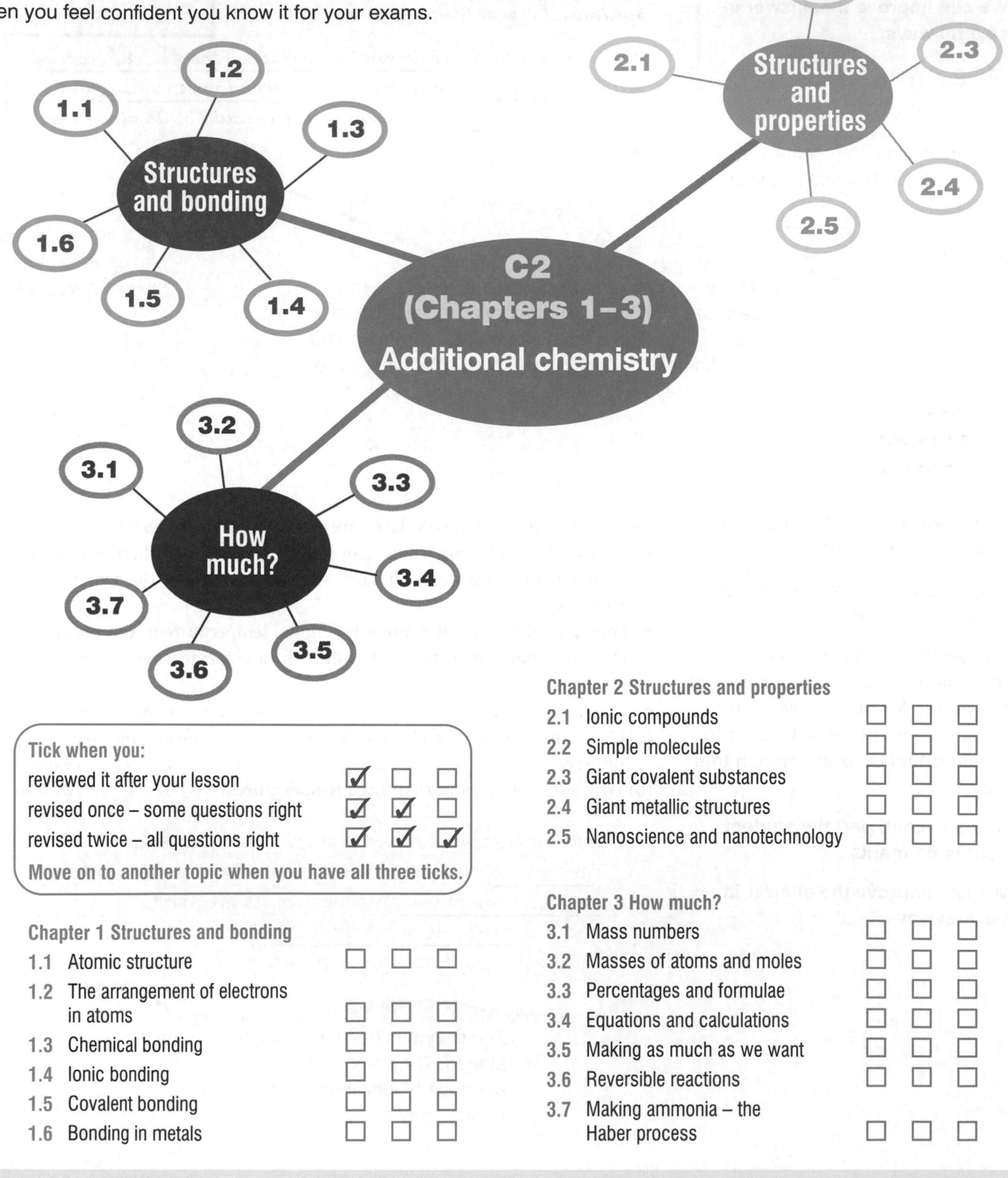

Tick when you:

reviewed it after your lesson ☑ ☐ ☐

revised once – some questions right ☑ ☑ ☐

revised twice – all questions right ☑ ☑ ☑

Move on to another topic when you have all three ticks.

Chapter 1 Structures and bonding

1.1	Atomic structure	☐	☐	☐
1.2	The arrangement of electrons in atoms	☐	☐	☐
1.3	Chemical bonding	☐	☐	☐
1.4	Ionic bonding	☐	☐	☐
1.5	Covalent bonding	☐	☐	☐
1.6	Bonding in metals	☐	☐	☐

Chapter 2 Structures and properties

2.1	Ionic compounds	☐	☐	☐
2.2	Simple molecules	☐	☐	☐
2.3	Giant covalent substances	☐	☐	☐
2.4	Giant metallic structures	☐	☐	☐
2.5	Nanoscience and nanotechnology	☐	☐	☐

Chapter 3 How much?

3.1	Mass numbers	☐	☐	☐
3.2	Masses of atoms and moles	☐	☐	☐
3.3	Percentages and formulae	☐	☐	☐
3.4	Equations and calculations	☐	☐	☐
3.5	Making as much as we want	☐	☐	☐
3.6	Reversible reactions	☐	☐	☐
3.7	Making ammonia – the Haber process	☐	☐	☐

What are you expected to know?

Chapter 1 Structures and bonding See students' book pages 88–103

- Atoms have a tiny central nucleus made of protons with a positive charge and neutrons with no charge. The nucleus is surrounded by electrons that have a negative charge, and are equal in number to the protons.
- All atoms of an element have the same number of protons. This is the atomic number (proton number) of the element, and is the order in which elements are arranged in the modern periodic table.
- Electrons are arranged in energy levels (shells). The pattern can be represented by numbers, e.g. 2,8,1 for sodium, or by dot and cross diagrams.
- Elements in the same group in the periodic table have the same number of electrons in their highest energy level (outer shell) and so they have similar chemical properties.
- Ions are formed when atoms lose or gain electrons. Ionic compounds are held together by strong attractions between oppositely charged ions in a giant structure.
- Covalent bonds are formed when pairs of electrons are shared between atoms, and these substances form molecules.
- Metals have giant structures of atoms.
- The atoms (or positively charged ions) are held together by delocalised electrons. This allows metals to conduct heat and electricity. [Higher Tier only]
- The layers of atoms in metals can slide over each other, allowing them to be bent and shaped.

Chapter 2 Structures and properties See students' book pages 104–115

- Ionic substances have high melting and boiling points. They conduct electricity when molten or in solution, but not when they are solids.
- Substances made of simple molecules have low melting and boiling points.
- This is because they have weak intermolecular forces. [Higher Tier only]
- The molecules have no charges and so do not conduct electricity.
- Atoms, such as carbon and silicon, that form several covalent bonds can form giant covalent structures.
- Nanoscience involves very small particles made of only a few hundred atoms, which gives these materials special properties and new uses.

Chapter 3 How much? See students' book pages 116–133

- Protons and neutrons have a mass of one unit, but electrons have very little mass. The mass number of an atom is the number of protons plus neutrons in the atom.
- Isotopes are atoms of the same element with different mass numbers.
- The relative atomic mass (A_r) of an element is the average mass of its isotopes compared with an atom of the ^{12}C isotope. [Higher Tier only]
- The relative formula mass of a compound (M_r) is found by adding up the relative atomic masses of the atoms shown in its formula.
- One mole of a substance is its A_r or M_r weighed out in grams.
- Percentages of elements in compounds can be calculated from A_r and M_r.
- Percentages by mass can also be used to calculate empirical formulae. [Higher Tier only]
- Masses of reactants and products can be calculated from balanced equations. [Higher Tier only]
- The percentage yield of a reaction does not always equal the theoretical yield. [Higher Tier only]
- The atom economy of a reaction is the mass of the atoms in the useful products as a percentage of the mass of all the atoms in the reactants. [Higher Tier only]
- Reversible reactions can be used efficiently in industrial processes like the Haber process, in which nitrogen and hydrogen react to make ammonia.
- In a closed system a reversible reaction can reach equilibrium. [Higher Tier only]

C2 1 Pre Test: Structures and bonding

1. What are the particles in an atom and in which part of an atom are they found?
2. What is the atomic number of an element?
3. How are electrons arranged in atoms?
4. What is special about the electron arrangement of elements in the same group of the periodic table?
5. What is special about the electron arrangement of the noble gases?
6. What happens to electrons when atoms of elements react?
7. What are ionic bonds?
8. Why does sodium chloride have the formula NaCl?
9. What is a covalent bond?
10. Draw a diagram to show the covalent bonds in a molecule of water.
11. How are atoms arranged in metals?
12. How are the atoms in metals held in position within their giant structures? [Higher Tier only]

students' book page 88

C2 1.1 Atomic structure

KEY POINTS

1. The nucleus of an atom is made of protons and neutrons.
2. Protons have a positive charge, electrons a negative charge and neutrons are not charged.
3. The atomic number (or proton number) of an element is the number of protons in the nucleus of its atoms.
4. Elements are arranged in order of their atomic numbers in the periodic table.

The nucleus at the centre of an atom contains two types of particle, called protons and neutrons. Protons have a positive charge and neutrons have no charge. Electrons are negatively charged particles that move around the nucleus. An atom has no overall charge, because the number of electrons is equal to the number of protons and their charges are equal and opposite.

All atoms of an element contain the same number of protons. This number is called the atomic number (or proton number) of the element. Elements are arranged in order of their atomic numbers in the periodic table. The atomic number tells you the number of protons and the number of electrons in atoms of the element.

Key words: protons, neutrons, electrons, atomic number, proton number, periodic table

AQA EXAMINER SAYS...

You can find the atomic number of an element in the periodic table and it tells you the number of protons and the number of electrons in atoms of the element.

CHECK YOURSELF

1. Name the three types of particle in atoms.
2. What are the charges on the three particles?
3. How many electrons are there in an atom of magnesium?

students' book page 90

C2 1.2 The arrangement of electrons in atoms

KEY POINTS

1 Electrons in atoms are in energy levels that can be represented by shells.
2 Electrons in the lowest energy level are in the shell closest to the nucleus.
3 Electrons occupy energy levels from the lowest first.
4 All the elements in a group of the periodic table have the same number of electrons in their highest energy level (outer shell).

GET IT RIGHT!

Only the first two noble gases have completely full outer shells, but the next energy level begins to fill after each noble gas.

Each electron in an atom is in an energy level. Energy levels can be represented as shells, with electrons in the lowest energy level closest to the nucleus. We can draw them as circles on a diagram, with electrons represented by dots or crosses.

The lowest energy level or first shell can hold two electrons, and the second energy level can hold eight. Electrons occupy the lowest possible energy levels, so the electronic structure of neon with 10 electrons is 2,8. Sodium with 11 electrons has an electronic structure of 2,8,1.

Elements in the same group of the periodic table have the same number of electrons in their highest energy level. All the elements in Group 1 have one electron in their highest energy level, showing that after each noble gas the next energy level begins to fill.

Key words: energy level, shell, electronic structure

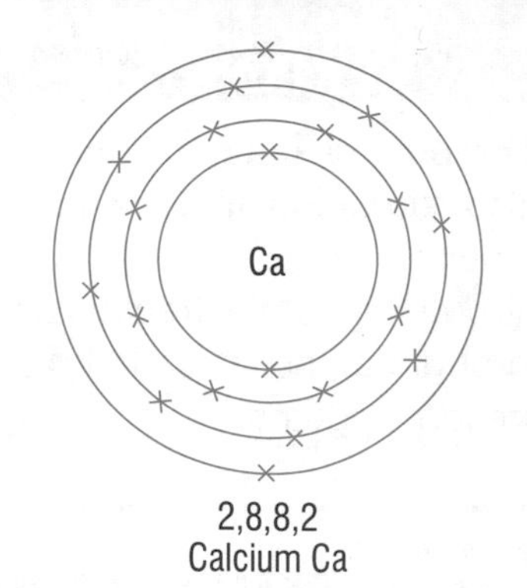

2,8,8,2
Calcium Ca

Once you know the pattern, you should be able to draw the energy levels of the electrons in any of the first 20 atoms (given their atomic number)

CHECK YOURSELF

1 What are electron shells?
2 Draw a diagram to show the electron arrangement of carbon.
3 Write the electronic structures of lithium, nitrogen and magnesium in number form.

students' book page 92

C2 1.3 Chemical bonding

KEY POINTS

1 Compounds are substances in which elements are chemically combined.
2 When elements react their atoms achieve stable arrangements of electrons.
3 Atoms gain or lose electrons to form ions or share electrons to form covalent bonds.

The noble gases are unreactive because their atoms have stable arrangements of electrons. Atoms of other elements can achieve stable electronic structures by gaining or losing electrons to form ions, or by sharing electrons to form covalent bonds. When an element in Group 1 reacts with an element in Group 7, an electron is transferred between atoms to form ions with the electron structure of a noble gas. The atoms of elements in Group 1 lose their single outer electron, for example sodium Na (2,8,1) forms sodium ions, Na^+ (2,8). The atoms of elements in Group 7 gain one electron to form ions, for example chlorine Cl (2,8,7) forms chloride ions, Cl^- (2,8,8). We can show this transferring of electrons using dot and cross diagrams. Positive and negative ions attract and form ionic bonds.

Key words: ions, covalent bonds, ionic bonds, sharing, transferring

EXAMINER SAYS…

Ionic bonds hold compounds made of ions together. Other compounds are held together by covalent bonds.

CHECK YOURSELF

1 Why are the noble gases unreactive?
2 Write the formula and electron arrangement for each of: a potassium ion, a magnesium ion, and an oxide ion.
3 Draw a dot and cross diagram to show the formation of lithium fluoride.

students' book page 94

C2 1.4 Ionic bonding

KEY POINTS

1. Compounds made of ions have giant structures that are very regular.
2. Strong forces of attraction act throughout the lattice to hold the ions together.

EXAMINER SAYS…

Many students seem to think that ionic compounds are made of pairs of ions. This is incorrect because ionic compounds have giant structures. The formula of an ionic compound is the simplest ratio of ions in the compound, and does not represent a molecule.

Ionic bonding holds oppositely charged ions together in giant structures. Strong electrostatic forces of attraction act in all directions. Each ion in the lattice is surrounded by ions with the opposite charge and so is held firmly in place.

The sodium chloride structure contains equal numbers of sodium and chloride ions as shown by its formula, NaCl. The sodium ions and chloride ions alternate to form a cubic lattice.

The ratio of ions in the formula and the structure of an ionic compound depend on the charges on the ions. For example, magnesium ions are Mg^{2+}, and chloride ions are Cl^- so the formula of magnesium chloride is $MgCl_2$. Its structure contains twice as many chloride ions as magnesium ions.

Key words: giant structure, lattice, formula

CHECK YOURSELF

1. What forces hold ions together in ionic bonding?
2. What does the formula NaCl tell you about sodium chloride?
3. What is the formula of calcium fluoride?

students' book page 96

C2 1.5 Covalent bonding

KEY POINTS

1. A covalent bond is formed when two atoms share a pair of electrons.
2. The number of covalent bonds an atom forms depends on the number of electrons it needs to achieve a stable electron arrangement.
3. Many covalently bonded substances consist of small molecules, but some have giant structures.

EXAMINER SAYS…

Covalent bonds join atoms together to form molecules. You should only use the word molecule when describing covalently bonded substances.

The atoms of non-metals need to gain electrons to achieve stable arrangements of electrons. They can do this by sharing electrons with other atoms. Each shared pair of electrons strongly attracts the two atoms, forming a 'covalent bond'.

Atoms of elements in Group 7 need to gain one electron and so form a single covalent bond. Those in Group 6 need to gain two electrons and form two covalent bonds. Atoms of elements in Group 5 can form three bonds and those in Group 4 can form four bonds.

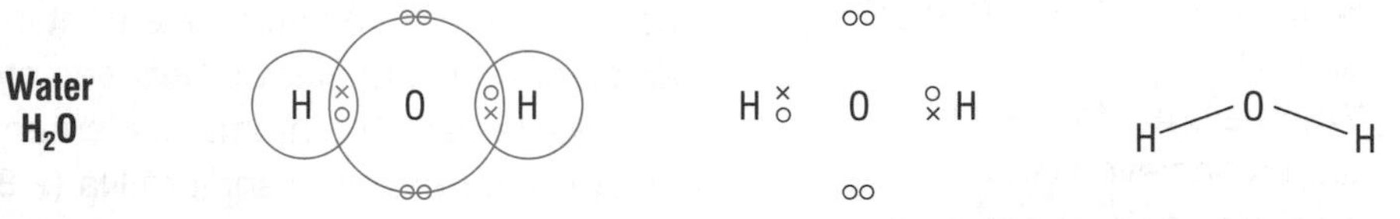

By choosing how we represent a covalent compound, we can show the outer energy level, the shared electrons or just the fact that there are a certain number of covalent bonds

Covalent bonds act only between the two atoms they bond, and so many covalently bonded substances consist of small molecules. Some atoms that can form several bonds, like carbon, can join together in giant covalent structures.

Key words: sharing, covalent bond, molecules, giant covalent structures

CHECK YOURSELF

1. What is a covalent bond?
2. How many covalent bonds can a silicon atom form?
3. Draw a dot and cross diagram for a molecule of ammonia.

students' book page 98

C2 1.6 Bonding in metals

KEY POINTS

1 Metals have giant structures of layers of atoms arranged in a regular pattern.
2 The electrons in the highest energy level delocalise. This results in strong electrostatic forces between these electrons and the positively charged metal ions, holding the metal together. [Higher Tier only]

The atoms in a metallic element are all the same size. They form giant structures in which layers of atoms are arranged in regular patterns. Although you cannot see individual atoms, you can see metal crystals on the surfaces of some metals. You can also grow metal crystals by displacement reactions. You can make models of metal structures by putting lots of small spheres like marbles together.

HIGHER

When metal atoms pack together the electrons in the highest energy level (the outer electrons) delocalise and move from one atom to another. This produces positive ions in a 'sea' of moving electrons. The delocalised electrons strongly attract the positive ions and hold the structure together.

Key words: crystals, delocalise

EXAMINER SAYS...

Some students do not make it clear in their answers that metallic bonding is strong and involves electrostatic forces.

CHECK YOURSELF

1 How are the atoms arranged in metal crystals?
2 Where can you see metal crystals?
3 What are delocalised electrons? [Higher Tier only]
4 What forces hold metal atoms in place in their giant structures? [Higher Tier only]

C2 1 End of chapter questions

1 **Why are the numbers of protons and electrons equal in an atom?**
2 **How many protons and electrons are in an atom of fluorine?**
3 **What is the arrangement of electrons in an atom of potassium?**
4 **What is special about the arrangement of electrons of the elements in Group 1?**
5 **What do we mean by a 'noble gas'?**
6 **Explain what happens when a sodium atom reacts with a fluorine atom.**
7 **What holds the ions together in an ionic lattice?**
8 **Potassium chloride is an ionic compound with formula KCl. What does its formula tell you about the structure of potassium chloride?**
9 **How is the number of covalent bonds that an atom can form related to its group in the periodic table?**
10 **Draw a diagram to show the bonds in hydrogen sulphide, H_2S.**
11 **How are the atoms arranged in a metal crystal?**
12 **How are the atoms held in position in a metal's giant structure? [Higher Tier only]**

C2 2 Pre Test: Structures and properties

1. **Why are ionic compounds always solid at room temperature?**
2. **When can ionic compounds conduct electricity?**
3. **Why are many covalent substances gases or liquids at room temperature?**
4. **Why do covalent compounds not conduct electricity?**
5. **Why can some covalently bonded substances form giant structures?**
6. **Why are diamond and graphite so different?**
7. **What happens to the atoms when a metal bends?**
8. **Do all metals conduct heat and electricity?**
9. **What is nanoscience?**
10. **What use can we make of nanoparticles?**

students' book page 104

C2 2.1 Ionic compounds

KEY POINTS

1. Ionic compounds have high melting and boiling points.
2. Ionic compounds conduct electricity when molten or in solution.

Ionic compounds have giant structures in which many strong electrostatic forces hold the ions tightly together. This means they are solids at room temperature. A lot of energy is needed to overcome the ionic bonds to melt the solids and so ionic compounds have high melting points and high boiling points.

However, when they have been melted the ions are free to move. This allows them to carry electrical charge, so the liquids conduct electricity. Some ionic solids dissolve in water because water molecules can split up the lattice. The ions are free to move in the solutions and so they also conduct electricity.

Key words: giant structures, ionic bonds, conduct, dissolve

EXAMINER SAYS...

There are many strong electrostatic forces of attraction to overcome to melt an ionic solid.

GET IT RIGHT!

Solid ionic compounds cannot conduct electricity.

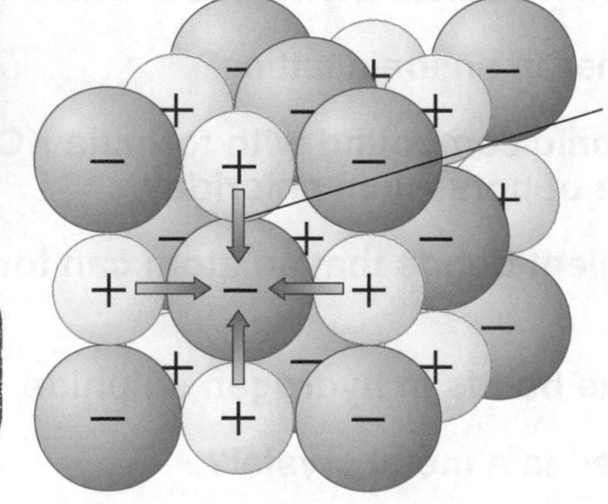

The attractive forces between the oppositely charged ions in an ionic compound are very strong

CHECK YOURSELF

1. Why do ionic solids have high melting points?
2. Why can some ionic solids dissolve in water?
3. Why can molten ionic substances conduct electricity?

students' book page 106

C2 2.2 Simple molecules

KEY POINTS

1 The forces between simple molecules are weak so many of these substances are gases or liquids at room temperature.
2 Simple molecules do not have a charge and so cannot conduct electricity.

The atoms in molecules are held together by strong covalent bonds. These bonds act only between the atoms within the molecule, and so simple molecules have little attraction for each other. Substances made of simple molecules have relatively low melting points and boiling points.

HIGHER

The forces of attraction between molecules, called 'intermolecular forces', are weak.

These forces are overcome when a molecular substance melts or boils. This means that substances made of small molecules have low melting and boiling points. Those with the smallest molecules, like H_2, Cl_2 and CH_4, have the weakest intermolecular forces and are gases at room temperature.

Larger molecules have stronger attractions and so may be liquids at room temperature, like Br_2 and C_6H_{14}, or solids with low melting points, like I_2.

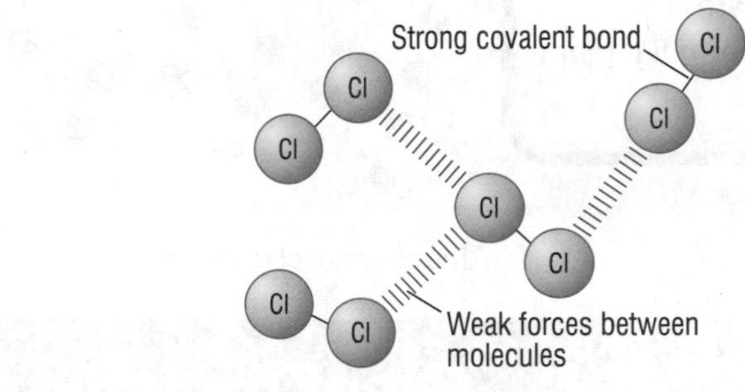

There are two types of attraction in molecular substances

Many students, when trying to explain melting or boiling, refer to bonds breaking. This suggests they think the covalent bonds are breaking when molecular substances melt, which is not correct. You should make it clear in your answers which forces you are writing about – covalent bonds or intermolecular forces.

GET IT RIGHT!

Covalent bonds are strong and difficult to break; intermolecular forces are much weaker and more easily overcome.

Key words: intermolecular forces

Molecules have no overall charge and cannot carry an electric current, so these substances do not conduct electricity.

CHECK YOURSELF

1 Why is oxygen, O_2, a gas at room temperature?

2 Why does petrol not conduct electricity?

3 What type of forces act between molecules? [Higher Tier only]

students' book page 108

C2 2.3 Giant covalent substances

KEY POINTS

1 Some covalently bonded substances form giant structures.
2 These substances have very high melting points.
3 Diamond and graphite are both forms of carbon but have many different properties.

Atoms of some elements can form several covalent bonds. These atoms can join together in giant covalent structures (sometimes called 'macromolecules'). Every atom in the structure is joined to several other atoms by strong covalent bonds. It takes an enormous amount of energy to break down the lattice and so these substances have very high melting points.

Diamond (a form of carbon) and silica (silicon dioxide) have regular three-dimensional giant structures and so they are very hard and transparent.

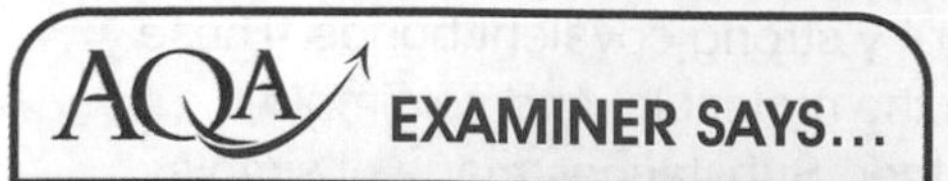

C2 2.3 Giant covalent substances (cont.)

students' book page 108

AQA EXAMINER SAYS...

Both diamond and graphite are macromolecules and so have high melting points. You should be able to explain the differences between the properties of diamond and graphite using the differences in their structures.

GET IT RIGHT!

In graphite the atoms are covalently bonded to form flat giant molecules.

Graphite is a form of carbon in which the atoms join in flat two-dimensional layers. There are only weak forces between the layers and so they slide over each other, making graphite slippery and grey.

Key words: macromolecules

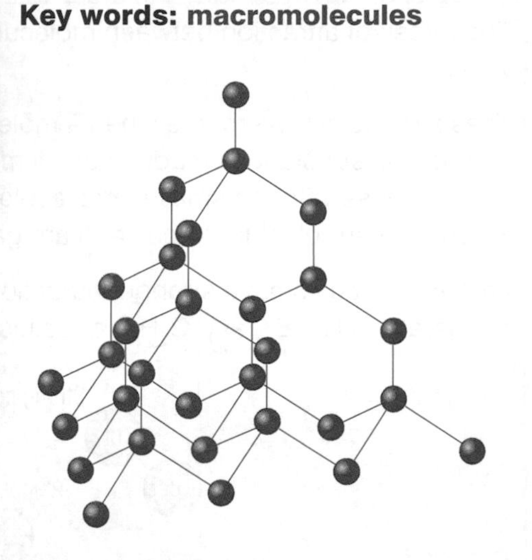

The structure of diamond

Graphite has delocalised electrons (as in a metal structure) along its layers and so conducts electricity.

CHECK YOURSELF

1. What is a macromolecule?
2. Why are diamond and silica transparent and hard?
3. Why is graphite slippery?
4. Why can graphite conduct electricity? [Higher Tier only]

C2 2.4 Giant metallic structures

students' book page 110

KEY POINTS

1. Metals can be bent and shaped because their layers of atoms can slide over each other.
2. Delocalised electrons move throughout metals and can carry heat and electricity. [Higher Tier only]

Metal atoms are arranged in layers. When a force is applied the layers of atoms can slide over each other. They can move into a new position without breaking apart, so the metal bends or stretches into a new shape. This means that metals are useful for making wires, rods and sheet materials.

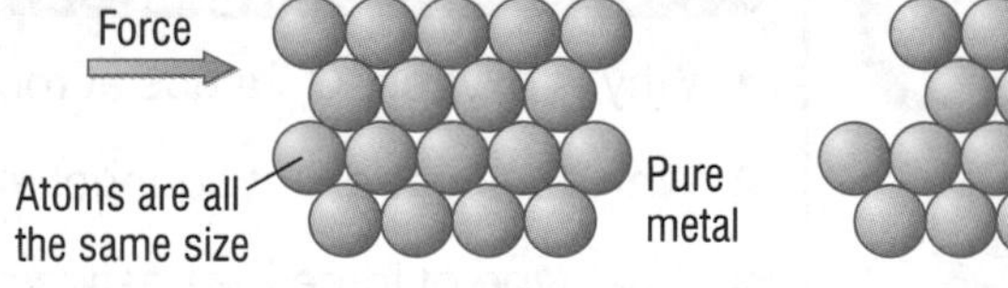

When a force is applied to a metal, the layers slide over each other

AQA EXAMINER SAYS...

Delocalised electrons are the key to understanding the properties of metals.

HIGHER

Delocalised electrons hold the atoms in place. The delocalised electrons are free to move throughout the metal structure. This means that they can flow as an electric current without changing the metal. They can carry heat energy and so metals are also very good conductors of heat. Many uses of metals depend on their ability to conduct heat and electricity.

Key words: delocalised electrons

CHECK YOURSELF

1. Why can we change the shape of metals?
2. How do metals conduct electricity? [Higher Tier only]
3. What allows metals to conduct heat? [Higher Tier only]

students' book page 112

C2 2.5 Nanoscience and nanotechnology

KEY POINTS

1. Nanoscience is about structures that are a few nanometres in size.
2. Nanoparticles behave differently to the same materials in bulk.

'Nano' means one billionth of a metre, i.e. 10^{-9} m.

AQA EXAMINER SAYS…

You need to know the types of application of nanoparticles, but you do not have to remember details of specific examples. Questions on this topic will often require you to apply your understanding to information supplied in the question.

When atoms are arranged into very small particles they behave differently to ordinary materials made of the same atoms. A nanometre is one billionth of a metre (or 10^{-9} m) and nanoparticles are a few nanometres in size. They contain a few hundred atoms arranged in a particular way. Their structures and very small sizes give them new properties that can make them very useful materials.

Nanoparticles have very large surface areas, exposing many more atoms at their surface than normal materials. Electrons can move through them more easily than ordinary materials. They can be very sensitive to light, heat, pH, electricity and magnetism.

Nanotechnology uses nanoparticles as very selective sensors, highly efficient catalysts, new coatings and construction materials with special properties, and to make drugs more effective.

Key words: nanometre, nanoparticles, nanotechnology

CHECK YOURSELF

1. About how many atoms are there in a typical nanoparticle?
2. Why do nanoparticles have different properties to ordinary materials?
3. Suggest three ways in which nanotechnology is being used.

C2 2 End of chapter questions

1. **Why does it take a lot of energy to melt ionic compounds?**
2. **Why can solutions of ionic compounds conduct electricity?**
3. **What are 'intermolecular forces'? [Higher Tier only]**
4. **What happens to the molecules when water boils? [Higher Tier only]**
5. **Why does diamond have a very high melting point?**
6. **How are the carbon atoms bonded in graphite?**
7. **Why can we pull metals into wires?**
8. **Why do metals stay the same when they conduct electricity? [Higher Tier only]**
9. **How are nanoparticles different to ordinary materials?**
10. **What is 'nanotechnology'?**

C2 3 Pre Test: How much?

1. What is the mass number of an atom?
2. What do we call atoms of the same element with different numbers of neutrons?
3. Why do we use relative atomic masses?
4. What is a mole?
5. How can we find the percentage of an element in a compound?
6. What is the difference between an empirical formula and a molecular formula? [Higher Tier only]
7. What do balanced equations tell us?
8. How much calcium oxide can we make from 10 g of calcium carbonate? [Higher Tier only]
9. What is meant by the yield of a reaction?
10. What do we mean by 'atom economy'?
11. How can we recognise a reversible reaction?
12. What is equilibrium?
13. How is ammonia manufactured?
14. Why is the yield of ammonia so low? [Higher Tier only]

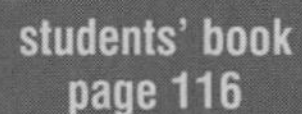

C2 3.1 Mass numbers

KEY POINTS

1. The mass of a proton is equal to the mass of a neutron.
2. The mass number of an atom is the total number of protons and neutrons in its nucleus.
3. Isotopes are atoms of the same element with different numbers of neutrons.

Protons and neutrons have equal masses. The relative masses of a proton and a neutron are both one unit. The mass of an electron is very small compared with a proton or neutron, and so the mass of an atom is made up almost entirely of its protons and neutrons. The total number of protons and neutrons in an atom is called its 'mass number'.

Atoms of the same element all have the same atomic number. The number of protons and electrons in an atom must always be the same, but there can be different numbers of neutrons. Atoms of the same element with different numbers of neutrons are called 'isotopes'. The number of neutrons in an atom is equal to its mass number minus its atomic number. We can show the mass number and atomic number of an atom like this:

$$^{23}_{11}Na$$

The number at the top (always larger, except in $^{1}_{1}H$) is the mass number.

Key words: mass number, isotopes

GET IT RIGHT!

number of neutrons =
mass number − atomic number

AQA EXAMINER SAYS...

Isotopes are atoms of the same element and have the same chemical properties. They have different physical properties because the different numbers of neutrons give them different masses. Some isotopes are also unstable and radioactive.

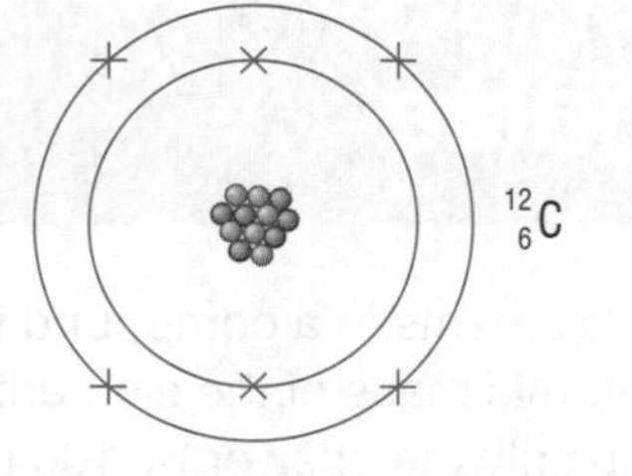

- Proton — Number of protons gives atomic number
- Neutron — Number of protons plus number of neutrons gives mass number

CHECK YOURSELF

1 What name do we use for the number of protons and neutrons in an atom?

2 What are isotopes?

3 Calculate the number of protons, neutrons and electrons in these atoms: $^{16}_{8}O$, $^{19}_{9}F$

students' book page 118

C2 3.2 Masses of atoms and moles

KEY POINTS

1 Relative atomic masses compare the masses of atoms.
2 The relative atomic mass of an element is an average value for the isotopes of an element. [Higher Tier only]
3 One mole of a substance is its relative formula mass in grams.

Atoms are much too small to weigh and so we use 'relative' atomic masses. These are often shown in periodic tables.

HIGHER

We use an atom of $^{12}_{6}C$ as a standard atom and compare the masses of all other atoms with this.

The relative atomic mass of an element (A_r) is an average value that depends on the isotopes the element contains. However, when rounded to a whole number it is often the same as the mass number of the main isotope of the element.

The relative formula mass (M_r) of a substance is found by adding up the relative atomic masses of the atoms in its formula. For example:

Worked example
Calculate the M_r of $CaCl_2$

Solution
A_r of Ca = 40, A_r of Cl = 35.5
so 40 + (35.5×2) = 111

The relative formula mass of a substance in grams is called 'one mole' of that substance. Using moles of substances allows us to calculate and weigh out in grams masses of substances with the same number of particles. One mole of sodium atoms contains the same number of atoms as one mole of chlorine atoms. For example:

Worked example
What is the mass of one mole of NaOH?

Solution
A_r of Na = 23, A_r of O = 16, A_r of H = 1
so 23 g + 16 g + g + 1 g = 40 g

Key words: relative atomic mass (A_r), relative formula mass (M_r), mole

AQA EXAMINER SAYS...

Moles are really useful because they give you a way of counting huge numbers of atoms, molecules and ions by weighing the substance in grams.

GET IT RIGHT!

One mole of a substance is its relative formula mass in grams.

EXAM HINTS

- Use the periodic table to look up relative atomic masses.
- When calculating relative formula masses, make sure you add up all the atomic masses in a formula correctly: H_2SO_4 has two hydrogen atoms, one sulfur atom and four oxygen atoms (A_r of H = 1, A_r of O = 16, A_r of S = 32), so its M_r is 98.

CHECK YOURSELF

1 Calculate the relative formula masses (M_r) of:
(a) H_2 (b) CH_4 (c) $MgCl_2$
(A_r of H = 1, A_r of C = 12, A_r of Mg = 24, A_r of Cl = 35.5)

2 What is the mass of one mole of water?

3 Atoms of which isotope are used as the standard atoms for relative atomic masses? [Higher Tier only]

students' book page 120

C2 3.3 Percentages and formulae

KEY POINTS

1. The percentage mass of an element in a compound can be calculated from its A_r and the M_r of the compound.
2. The empirical formula of a compound can be calculated from its percentage composition.

We can calculate the percentage of any of the elements in a compound from the formula of the compound. Divide the relative atomic mass of the element by the relative formula mass of the compound and multiply the answer by 100 to convert it to a percentage. This can be useful when deciding if a compound is suitable for a particular purpose or to identify a compound.

For example:

Worked example

Find the percentage of carbon in carbon dioxide (A_r of C = 12, A_r of O = 16)

Solution

$$M_r \text{ of } CO_2 = 12 + (16 \times 2) = 44$$

$$\text{So percentage of carbon} = \left(\frac{12}{44}\right) \times 100 = 27.3\%$$

BUMP UP YOUR GRADE

Grade C students should be able to calculate the percentage of an element in a compound. For top grades you should be able to calculate the empirical formula of a compound with two or three elements.

HIGHER

The empirical formula is the simplest ratio of the atoms or ions in the compound. It is the formula used for ionic compounds, but for covalent compounds it is not always the same as the molecular formula. For example, the molecular formula of ethane is C_2H_6, but its empirical formula is CH_3.

We can calculate the empirical formula of a compound from its percentage composition:

- Divide the mass of each element in 100 g of the compound by its A_r to give the ratio of atoms.
- Then convert this to the simplest whole number ratio.

For example:

Worked example

Work out the empirical formula of the hydrocarbon that contains 80% carbon.

Solution

100 g of hydrocarbon contains 80 g of C and 20 g of H.

$$\text{Number of moles of carbon} = \frac{80}{12} = 6.67$$

$$\text{Number of moles of hydrogen} = \frac{20}{1} = 20$$

Ratio of atoms is 6.67 C : 20 H

Simplest ratio is 1 C : 3 H

So empirical formula is CH_3.

AQA EXAMINER SAYS...

Only higher attaining students will be expected to calculate empirical formulae.

Key words: empirical formula, molecular formula

CHECK YOURSELF

1. What is the percentage by mass of calcium in calcium oxide, CaO?
2. What is the empirical formula of propene, C_3H_6? [Higher Tier only]
3. A compound of iron and chlorine contains 44% iron by mass. What is its empirical formula? [Higher Tier only]

students' book page 122

C2 3.4 Equations and calculations

KEY POINTS

1. Balanced chemical equations can be used to calculate masses of reactants and products.
2. In an equation, $2Cl_2$ can mean 2 molecules of chlorine or two moles of chlorine molecules.

You can work in moles or you can use the relative masses in the equation when doing calculations, but don't forget to give correct units in your answer.

EXAM HINTS

Use the periodic table on your Data Sheet (see page 113) to find relative atomic masses if they are not given in the question.

HIGHER

Chemical equations show the reactants and products of a reaction. When they are balanced they show the amounts of atoms, molecules or ions in the reaction. For example:

$$2Mg + O_2 \rightarrow 2MgO$$

shows that two atoms of magnesium react with one molecule of oxygen to form two magnesium ions and two oxide ions.

Working in relative masses this becomes:

$$(2 \times A_r \text{ of Mg}) + (2 \times A_r \text{ of O}) \text{ gives } (2 \times M_r \text{ of MgO}) \text{ or } (2 \times 24 + 2 \times 16 = 2 \times 40)$$

If we work in moles, the equation tells us that two moles of magnesium atoms react with one mole of oxygen molecules to produce two moles of magnesium oxide.

This means 48 g of magnesium react with 32 g of oxygen to give 80 g of magnesium oxide. (A_r of Mg = 24, A_r of O = 16)

If we have a known mass of magnesium, say 5 g, we can work out the mass of magnesium oxide using moles.

In this case 5 g = $\frac{5}{24}$ moles of magnesium and so it will produce:

$$\frac{5}{24} \times 40\,\text{g} = 8.33\,\text{g of MgO}$$

We can also do it by calculating the proportion of the amounts in the equation:

$$5\,\text{g Mg will produce } 5 \times \frac{80}{48}\,\text{g} = 8.33\,\text{g MgO}$$

CHECK YOURSELF

1. Balance this equation: $H_2 + Cl_2 \rightarrow HCl$
2. Calculate the mass of sodium chloride that can be made from one mole of sodium in the reaction: $2Na + Cl_2 \rightarrow 2NaCl$
3. Calculate the mass of copper oxide that can be made from 10 g of copper in the reaction: $2Cu + O_2 \rightarrow 2CuO$

students' book page 124

C2 3.5 Making as much as we want

KEY POINTS

1. The yield of a reaction compares the amount of product actually made with the maximum amount that could be made.
2. Atom economy measures how much of the starting materials becomes useful products.

The yield of a chemical process compares how much you actually make with the maximum amount possible. When you actually carry out chemical reactions it is not possible to collect the amounts calculated from the chemical equations. Reactions may not go to completion and some product may be lost in the process.

HIGHER

The yield is often calculated as a percentage:

$$\text{percentage yield} = \frac{\text{amount of product collected}}{\text{maximum amount of product possible}} \times 100$$

students' book page 124

C2 3.5 Making as much as we want (cont.)

You should be able to comment on given values for yield and atom economy for different chemical processes. In more difficult questions you may be given data to calculate these quantities.

CHECK YOURSELF

1. Why is it not usually possible to collect the maximum yield from a reaction?
2. 5.2 g of potassium chloride, KCl, was made from 5.6 g of potassium hydroxide, KOH. What was the percentage yield? [Higher Tier only]
3. Quicklime is produced from limestone by the reaction: $CaCO_3 \rightarrow CaO + CO_2$ What is the percentage atom economy of this process? [Higher Tier only]

Atom economy

It is also important to consider the amount of the starting materials that ends up in the useful products. This is called the 'atom economy' of a process.

HIGHER

Atom economy is calculated by finding the mass of all the atoms in the starting materials and comparing this with the mass of the atoms in the useful product. It is also often worked out as a percentage:

$$\text{percentage atom economy} = \frac{\text{relative formula mass of useful product}}{\text{relative formula mass of all products}} \times 100$$

Worked example

Zinc is extracted by heating its oxide with carbon. Carbon monoxide is also produced in the reaction.

$$ZnO + C \rightarrow Zn + CO$$

Work out the atom economy from the equation shown above.
(A_r of Zn = 65, A_r of C = 12, A_r of O = 16)

Solution

$$\text{atom economy} = \frac{\text{relative formula mass of useful product (Zn)}}{\text{relative formula mass of all products (Zn + CO)}} \times 100$$

$$= \frac{65}{65 + (12 + 16)} = \frac{65}{93} = 70\%$$

- To avoid waste both percentage yield and atom economy should be as high as possible.

Key words: yield, atom economy

students' book page 126

C2 3.6 Reversible reactions

KEY POINTS

1. Reversible reactions go in both directions.
2. Reversible reactions can reach equilibrium in closed systems. [Higher Tier only]
3. Changing the conditions can change the amounts of reactants and products. [Higher Tier only]

CHECK YOURSELF

1. What is a reversible reaction?
2. What can happen to a reversible reaction in a closed system? [Higher Tier only]
3. How can the amounts of reactants and products in a reaction mixture in a closed system be changed? [Higher Tier only]

If the products of a chemical reaction can react to produce the reactants the reaction can go in both directions. This type of reaction is called a reversible reaction and is represented with the symbol $\rightleftharpoons$

HIGHER

When there are no products the reaction can only go in the forward direction, but as products build up the reverse reaction can happen. In a closed system nothing can escape and the rates of both forward and backward reactions will become equal. When this happens the system is at equilibrium.

If the conditions of the system are changed the amounts of reactants and products may change. Increasing the concentration of a substance will increase the rate of the reaction away from that substance. If the system is open and products can escape the forward reaction will continue to completion.

1) A+B →	(Reactants only at start of reaction)
2) A+B ⇌ C+D	(Rate of → much greater than ← at first)
3) A+B ⇌ C+D	(Rate of ← increases as C+D build up. Rate of → slows down as reactants get used up)
4) A+B ⇌ C+D	(Eventually the rates of → and ← are the same)

A reversible reaction

Key words: reversible, closed system, equilibrium

students' book page 128

C2 3.7 Making ammonia – the Haber process

KEY POINTS

1 The Haber process produces ammonia from nitrogen and hydrogen.
2 The reaction uses a high temperature, high pressure and a catalyst to produce a reasonable yield of ammonia in a short time.

The Haber process is used to manufacture ammonia, which can be used to make fertilisers and other chemicals.

Nitrogen from the air and hydrogen, which is usually obtained from natural gas, are purified and mixed in the correct proportions. The gases are passed over an iron catalyst at a temperature of about 450°C and a pressure of about 200 atmospheres.

The reaction is reversible and so some ammonia breaks down into nitrogen and hydrogen. The gases are cooled so the ammonia condenses. The liquid ammonia is removed from the unreacted gases and they are recycled.

The yield is less than 20%, but the ammonia is produced quickly and no gases are wasted.

AQA EXAMINER SAYS...

Conditions in the Haber process can vary in different places but are always chosen to produce the best yield as quickly as possible.

CHECK YOURSELF

1 What are the raw materials used to make ammonia?
2 Write a word equation and a balanced equation for the reaction to make ammonia.
3 Why does the reaction not go to completion?

C2 3 End of chapter questions

1 **Why do we not include electrons in mass numbers?**
2 **How many protons, neutrons and electrons are there in an atom of $^{27}_{13}Al$?**
3 **What is the relative formula mass of Na_2O?**
4 **What is the mass of one mole of CO_2?**
5 **What is the percentage by mass of carbon in methane, CH_4?**
6 **What is the empirical formula of the compound of iron and oxygen that contains 70% iron? [Higher Tier only]**
7 **Balance this equation: $CH_4 + O_2 \rightarrow O_2 + H_2O$ [Higher Tier only]**
8 **Calculate the mass of lithium chloride you can make from 2.4 g of lithium hydroxide: $HCl + LiOH \rightarrow LiCl + H_2O$ [Higher Tier only]**
9 **Calculate the percentage yield if 4.1 g of zinc was made from 8.1 g of zinc oxide in the reaction: $ZnO + H_2 \rightarrow Zn + H_2O$ [Higher Tier only]**
10 **What is the percentage atom economy of the reaction to produce copper from copper oxide? $2CuO + C \rightarrow 2Cu + CO_2$ [Higher Tier only]**
11 **The thermal decomposition of ammonium chloride is a reversible reaction:**

$$NH_4Cl \rightleftharpoons NH_3 + HCl$$

Explain what this means.
12 **What sort of system is needed for an equilibrium? [Higher Tier only]**
13 **What conditions are used for the Haber process?**
14 **What happens to the unreacted gases in the process?**

EXAMINATION-STYLE QUESTIONS

1 Sodium atoms have 11 electrons and oxygen atoms have 8 electrons.

(a) Draw a dot and cross diagram to show the arrangement of electrons in a sodium atom. (2 marks)

(b) Draw a dot and cross diagram to show the arrangement of electrons and the charge for a sodium ion. (2 marks)

(c) Draw a dot and cross diagram to show the arrangement of electrons and the charge on an oxide ion. (2 marks)

(d) Sodium oxide is a solid with a high melting point. Explain why, in terms of its structure and bonding. (2 marks)

(e) Explain why the formula of sodium oxide is Na_2O. (2 marks)

2 (a) Complete the table about the particles in atoms:

Particle	Relative charge	Relative mass
proton	+1	(i)
neutron	(ii)	1
electron	(iii)	very small

(3 marks)

(b) Chlorine has two main isotopes, $^{35}_{17}Cl$ and $^{37}_{17}Cl$.

(i) How many protons, neutrons and electrons are there in an atom of $^{35}_{17}Cl$? (3 marks)

(ii) What is the difference between these two isotopes of chlorine? (1 mark)

(c) (i) The relative atomic mass of chlorine is 35.5. Why does it have this value? (2 marks)

(ii) Explain why these two isotopes of chlorine have the same chemical properties. (2 marks)

3 Ammonia has the formula NH_3. It is a gas at room temperature.

(a) Name the type of bonding in ammonia. (1 mark)

(b) Draw a dot and cross diagram to represent the bonding in ammonia. (2 marks)

(c) Explain, in terms of its structure, why ammonia is a gas at room temperature. (2 marks)
[Higher]

4 (a) Ammonium chloride, $NH_4Cl(s)$, decomposes when heated to form ammonia gas, $NH_3(g)$ and hydrogen chloride gas $HCl(g)$. When cooled the gases recombine to form ammonium chloride, so the reaction is reversible.

Write a word equation to show this reversible reaction. (2 marks)

(b) Ammonium chloride is used as a fertiliser.

(i) Calculate the mass of one mole of ammonium chloride. (2 marks)

(ii) Calculate the percentage by mass of nitrogen in ammonium chloride. (2 marks)

5 Germanium, Ge, is an element in Group 4 of the periodic table. It is a white, shiny, brittle solid with a very high melting point. It is used in the electronics industry because it conducts a small amount of electricity. It is made from germanium oxide, GeO_2, by reduction with hydrogen. Germanium oxide is an ionic solid and it reacts with hydrochloric acid to produce germanium chloride, $GeCl_4$. Germanium chloride is a volatile liquid. It has small molecules with covalent bonds.

(a) Write a word equation for the reduction of germanium oxide. (2 marks)

(b) Balance the equation for the reaction of germanium oxide with hydrochloric acid:

$$GeO_2 + HCl \rightarrow GeCl_4 + H_2O$$

(2 marks)

(c) Write the formulae of the ions present in germanium oxide. (2 marks)

(d) Draw a dot and cross diagram to show the bonding in germanium chloride. You need to show only the outer electrons. (2 marks)

(e) Germanium has some properties of a metal and some properties of a non-metal. What evidence is there in the information in this question that suggests:

(i) that germanium is a metal? (2 marks)

(ii) that germanium is a non-metal? (2 marks)

(f) Germanium could also be produced by reduction with carbon:

$$GeO_2 + C \rightarrow Ge + CO_2$$

(i) Calculate the percentage atom economy of this reaction. (2 marks)

(ii) Calculate the percentage atom economy for reduction with hydrogen. (2 marks)

(iii) Suggest *two* reasons why reduction with hydrogen is better than reduction with carbon. (2 marks)
[Higher]

The answer is worth 4 marks out of the 5 available. The responses worth a mark are underlined in red.

We can improve the answer in several ways:

Explain, as fully as you can, why sodium chloride has a high melting point. *(5 marks)*

Sodium chloride has a giant structure. It is made of sodium ions and chloride ions that have opposite charges. The ions attract each other in the lattice and there are lots of them in a crystal so lots of energy is needed to break them apart and melt them down.

The student has not expressed the ideas sufficiently well to gain all 5 marks. The important point about ionic bonding that has been missed is that the attractions are strong. Adding the word 'strongly' would have scored an additional mark.

This suggests that the student may think that ions are separated when they melt, rather than the forces being overcome sufficiently so that they can move about. Also, the ions themselves do not melt – it is the solid that melts.

The student has scored 4 marks out of a possible 6.

Note that part (c) would only appear on a Higher Tier exam paper.

The responses worth a mark are underlined in red.

We can improve the answer in several ways:

Ammonium nitrate, NH_4NO_3, is used as a fertiliser.

(a) Calculate the relative formula mass of ammonium nitrate, NH_4NO_3. *(2 marks)*

(b) Use your answer to part (a) to calculate the percentage by mass of nitrogen in ammonium nitrate. *(2 marks)*

(c) A student made 5.2 g of ammonium nitrate crystals by reacting 0.1 mole of nitric acid with ammonia. What was the percentage yield? *(2 marks)*

(a) $NH_4NO_3 = 14 + 4 + 14 + 16 \times 3$
$= 30 + 48 = 78$

(b) N + N $= 14 + 14 = 28$
$28 \times 100/78 = 35.9\%$

(c) 0.1 mole nitric acid → 0.1 mole ammonium nitrate
0.1 mole $NH_4NO_3 = 7.8$ g
% yield $= 5.2 \times 100/7.8 = 66\%$

The first part of the working is correct, but the answer is incorrect.

The student uses the answer from part (a) to calculate the percentage of nitrogen. The working is correct using the incorrect answer from (a) and so the error is carried forward and both marks can be awarded. (It is important to carry on working in any calculation using the answers you have written, even if you think your answer to any part may not be correct.)

The working is correct using the incorrect answer from (a), but the student has rounded the answer incorrectly. It should be 67% to two significant figures, or 66.7% to three figures.

C2 | Additional chemistry (Chapters 4–7)

Checklist

This spider diagram shows the topics in the second half of the unit. You can copy it out and add your notes and questions around it, or cross off each section when you feel confident you know it for your exams.

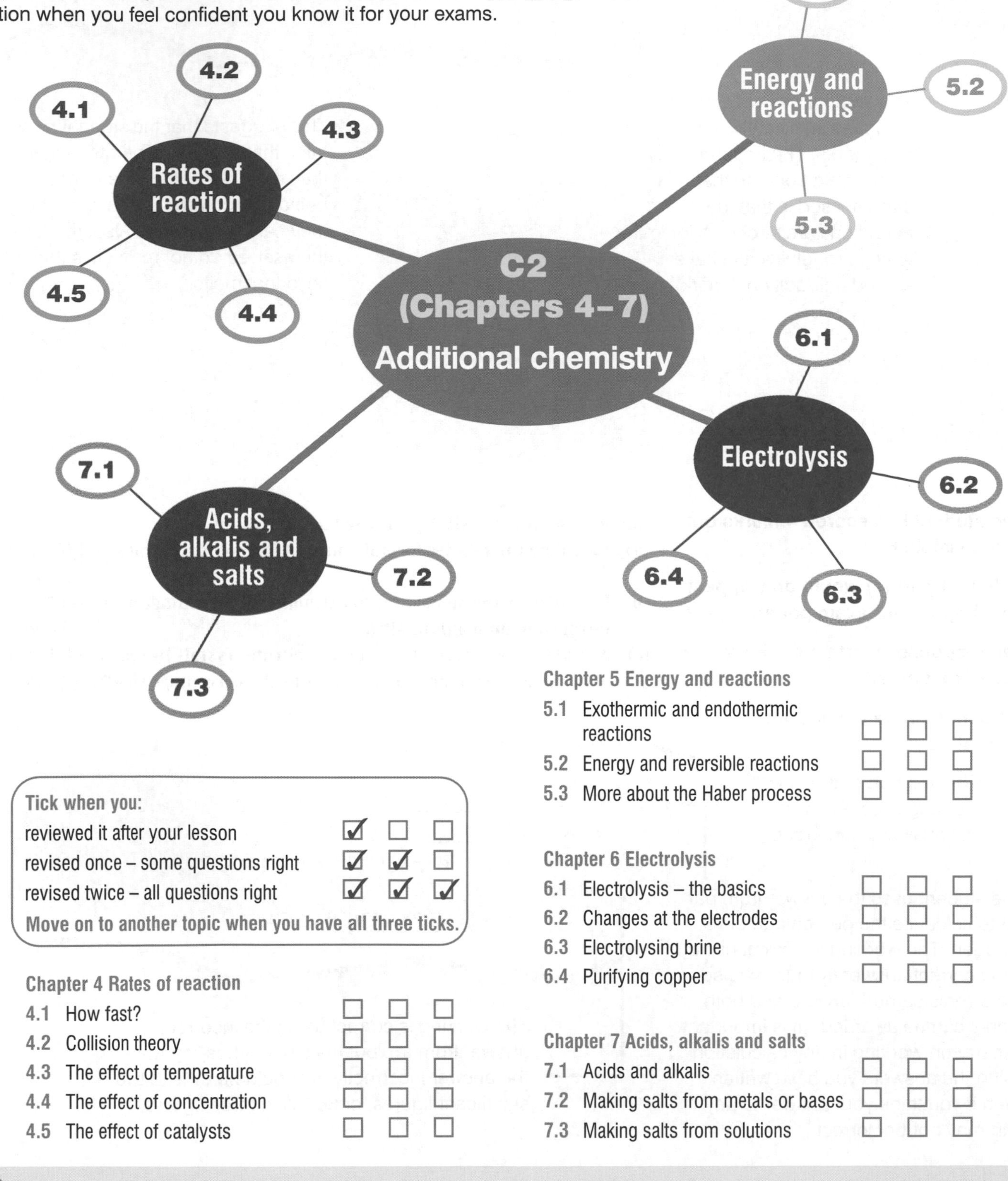

Tick when you:

reviewed it after your lesson ☑ ☐ ☐

revised once – some questions right ☑ ☑ ☐

revised twice – all questions right ☑ ☑ ☑

Move on to another topic when you have all three ticks.

Chapter 4 Rates of reaction

- 4.1 How fast? ☐ ☐ ☐
- 4.2 Collision theory ☐ ☐ ☐
- 4.3 The effect of temperature ☐ ☐ ☐
- 4.4 The effect of concentration ☐ ☐ ☐
- 4.5 The effect of catalysts ☐ ☐ ☐

Chapter 5 Energy and reactions

- 5.1 Exothermic and endothermic reactions ☐ ☐ ☐
- 5.2 Energy and reversible reactions ☐ ☐ ☐
- 5.3 More about the Haber process ☐ ☐ ☐

Chapter 6 Electrolysis

- 6.1 Electrolysis – the basics ☐ ☐ ☐
- 6.2 Changes at the electrodes ☐ ☐ ☐
- 6.3 Electrolysing brine ☐ ☐ ☐
- 6.4 Purifying copper ☐ ☐ ☐

Chapter 7 Acids, alkalis and salts

- 7.1 Acids and alkalis ☐ ☐ ☐
- 7.2 Making salts from metals or bases ☐ ☐ ☐
- 7.3 Making salts from solutions ☐ ☐ ☐

What are you expected to know?

Chapter 4 Rates of reaction See students' book pages 134–147

- The rate of a chemical reaction is found by measuring the amount of a reactant used or the amount of a product formed in a unit of time.
- Chemical reactions only happen when particles collide with enough energy to react. The minimum energy they need to react is called the 'activation energy'.
- Increasing temperature, concentration of solutions, pressure of gases, surface area of solids and using a catalyst increases the rate of reactions.
- Increasing the temperature of reactants causes the particles to collide more often and with more energy.
- Increasing the concentration of reactants, pressure of gases, or the surface area of solid reactants causes particles to collide more frequently.
- Concentrations of solutions measured in moles per dm^3 enable you to compare the number of particles of the substances dissolved in the solution. [Higher Tier only]
- Equal volumes of gases at the same temperature and pressure contain the same number of molecules (and the same number of moles). [Higher Tier only]
- Catalysts change the rate of particular reactions and are important in speeding up industrial processes.

Chapter 5 Energy and reactions See students' book pages 148–157

- Exothermic reactions transfer heat to the surroundings, and endothermic reactions take in heat from the surroundings.
- Reversible reactions are exothermic in one direction and endothermic in the other direction.
- If the temperature is raised, the yield from the endothermic reaction increases and the yield from the exothermic reaction decreases. [Higher Tier only]
- In reversible reactions involving gases, high pressure favours the reaction that produces the smallest number of molecules of gas. [Higher Tier only]
- It is important to minimise the energy used and the energy wasted in industrial processes for economic and environmental reasons.

Chapter 6 Electrolysis See students' book pages 158–169

- Ions in molten ionic compounds or in solutions are free to move and so can conduct electricity. The compounds are broken down into elements by electrolysis.
- Positive ions are attracted to the negative electrode where they gain electrons (reduction) and negative ions lose electrons at the positive electrode (oxidation).
- Reactions at electrodes depend on the ions that are present, their concentrations and what the electrode is made from.
- Electrolysis of sodium chloride solution is used in industry to produce chlorine, sodium hydroxide and hydrogen.
- Copper can be purified by electrolysis using copper electrodes in a solution containing copper ions.
- Reactions at electrodes can be represented by half equations. [Higher Tier only]

Chapter 7 Acids, alkalis and salts See students' book pages 170–181

- Insoluble salts can be made by mixing solutions of ions to form a precipitate.
- Soluble salts can be made from acids using a suitable metal, an insoluble base or an alkali. Then the solution can be crystallised to obtain the salt.
- Hydrogen ions $H^+(aq)$ make solutions acidic and hydroxide ions $OH^-(aq)$ make solutions alkaline. The pH scale measures the acidity or alkalinity of a solution.
- Neutralisation is the reaction of $H^+(aq)$ ions with $OH^-(aq)$ ions to produce $H_2O(l)$.

C2 4 Pre Test: Rates of reaction

1. **What is the rate of a reaction?**
2. **How can we measure the rate of a reaction?**
3. **What must happen to particles for them to react?**
4. **Why does changing the conditions change the rate of a reaction?**
5. **Why does increasing the temperature increase the rate of a reaction?**
6. **What happens to the rate of many reactions if you increase the temperature by 10°C?**
7. **What happens to the rate of a reaction if you increase the concentration of the reactants?**
8. **How does changing the pressure affect the rate of reactions between gases?**
9. **What is a catalyst?**
10. **Why are catalysts used in many industrial processes?**

students' book page 134

C2 4.1 How fast?

KEY POINTS

1. The rate of a reaction tells us how quickly reactants become products.
2. We can measure how quickly reactants are used up or how quickly products are formed.
3. Measuring a rate involves measuring an amount and the time it takes.

The rate of a reaction measures the speed of a reaction or how fast it is. The rate can be found by measuring how much of a reactant is used up or how much of a product is formed in a certain time.

An alternative way is to measure the time for a certain amount of reactant to be used or product to be formed.

Whichever way it is done it involves measuring both an amount and a time because:

$$\text{rate} = \frac{\text{amount of reactant used}}{\text{time}} \text{ OR } \frac{\text{amount of product formed}}{\text{time}}$$

The simplest measurements we can make are the mass of gas released or the volume of gas produced at intervals of time. Another method is to measure the time it takes for a certain amount of solid to appear in a solution. Other possible ways include measuring changes in the colour, concentration, or pH of a reaction mixture over time.

Key words: rate of reaction

EXAMINER SAYS…

The faster the rate, the shorter the time it takes for the reaction. So rate is inversely proportional to time.

CHECK YOURSELF

1. Explain what we mean by the 'rate of a reaction'.
2. What two things must we measure to find the rate of a reaction?
3. How can we measure the rate of a reaction that gives off a gas?

students' book page 136 — C2 4.2 Collision theory

KEY POINTS

1. Reactions happen when particles collide with enough energy to bring about a change.
2. Factors that affect the frequency or energy of collisions will change the rate of a reaction.

AQA EXAMINER SAYS...

When describing how rate is affected by collisions you should make it clear that it is the frequency of collisions that matters. In other words, more collisions in the same time will increase the rate.

The collision theory states that reactions can only happen if particles collide. However, just colliding is not enough. The particles must collide with enough energy to change into new substances. The minimum energy they need is called the 'activation energy'.

Factors that increase the chance of collisions or the energy of the particles will increase the rate of the reaction. Increasing the temperature, concentration of solutions, pressure of gases, surface area of solids and using a catalyst will increase the rate of a reaction.

Breaking large pieces of a solid into smaller pieces exposes new surfaces and so increases the surface area. This means there are more collisions in the same time and so a powder reacts faster than large lumps of a substance. There are many examples of powders reacting very rapidly.

Key words: collision theory, activation energy, surface area

GET IT RIGHT!

Particles must collide with more than the minimum energy needed for a reaction to happen.

CHECK YOURSELF

1. What is the collision theory of reactions?
2. What factors increase the rate of a reaction?
3. Why do powders react faster than solids?

students' book page 138 — C2 4.3 The effect of temperature

KEY POINTS

1. Increasing the temperature increases the rate of reactions.
2. A small increase in temperature produces a large change in reaction rates.

Increasing the temperature increases the speed of the particles in a reaction mixture. This means they collide more often, which increases the rate of reaction. As well as colliding more frequently they collide with more energy, which also increases the rate of reaction.

Therefore, a small change in temperature has a large effect on reaction rates. At ordinary temperatures a rise of 10°C will roughly double the rate of many reactions, so they go twice as fast. A decrease in temperature will slow reactions down, and a change of 10°C will double the time that many reactions take. This is why we refrigerate or freeze food so it stays fresh for longer.

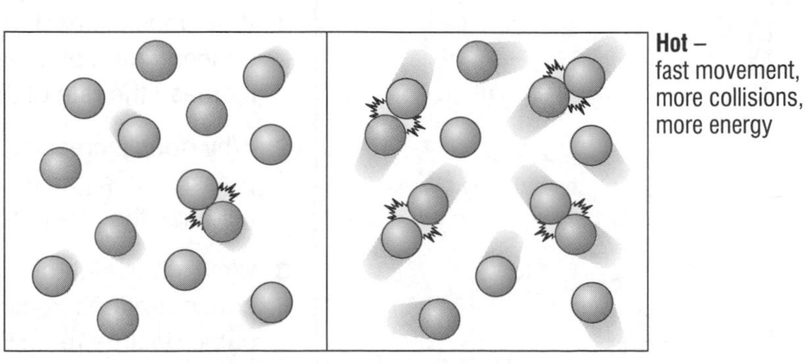

More collisions with more energy – both of these increase the rate of a chemical reaction as the temperature increases

CHECK YOURSELF

1. Why does a small change in temperature have a large effect on reaction rates?
2. What temperature change doubles the rate of many reactions?
3. Explain why refrigerating food makes it last longer.

students' book page 140

C2 4.4 The effect of concentration

KEY POINTS

1. Increasing the concentration of reactants increases the rate of reactions.
2. Concentrations of solutions are measured in moles per cubic decimetre (mol/dm^3). [Higher Tier only]
3. Equal volumes of gases at the same temperature and pressure contain equal numbers of molecules. [Higher Tier only]

AQA EXAMINER SAYS...

You will not have to work out the number of moles of substances in solutions or in gases in this unit. However, it is important that you understand that equal volumes of solutions of equal concentrations contain the same number of particles. You also need to know that if the concentration is doubled the number of particles in a given volume is doubled.

If the concentration of a solution is increased there are more particles dissolved in the same volume. This means the dissolved particles are closer together and so they collide more often. Increasing the concentration of a reactant therefore increases the rate of a reaction.

Low concentration/ low pressure

High concentration/ high pressure

Increasing concentration and pressure mean that particles are closer together. This increases the number of collisions between particles, so the reaction rate increases.

In a similar way, increasing the pressure of a gas puts more molecules into the same volume, and so they collide more frequently. This increases the rate of reactions that have gases as reactants.

HIGHER

Equal volumes of gases at the same temperature and pressure contain equal numbers of molecules. So 10 cm^3 of hydrogen contains the same number of molecules as 10 cm^3 of oxygen at the same temperature and pressure.

Measuring concentrations of solutions in moles per cubic decimetre (mol/dm^3) means we can measure out equal numbers of particles of the solutes by taking equal volumes of the same concentration.

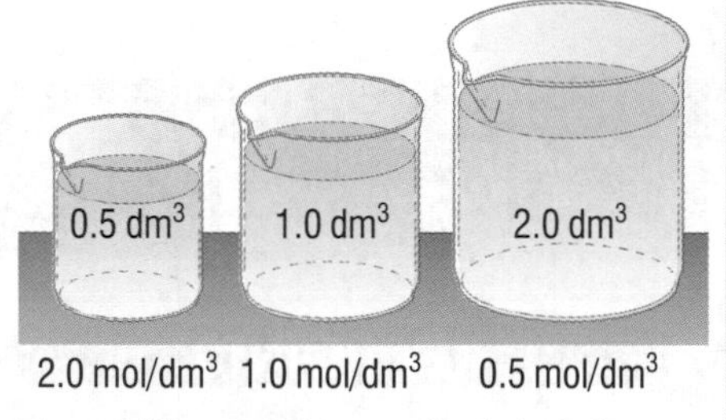

These different volumes of solution all contain the same amount of solute – but at different concentrations

The graph below shows how the mass of a reaction mixture changes as a gas is given off. The three lines are drawn for different concentrations of a solution reacting with an excess of solid reactant:

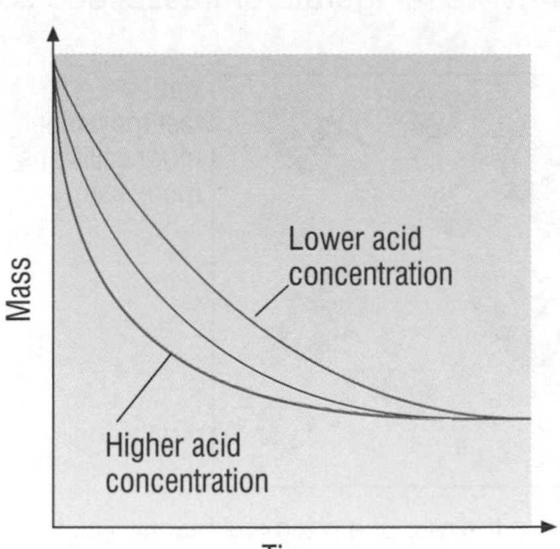

CHECK YOURSELF

1. Why does increasing the concentration of a reactant increase the rate of the reaction?
2. Why does increasing the pressure of a gas reactant increase the rate of the reaction?
3. Why is it useful to measure concentrations in moles per cubic decimetre? [Higher Tier only]

students' book page 142

C2 4.5 The effect of catalysts

KEY POINTS

1 Catalysts change the rates of chemical reactions but are not used up in the reaction.
2 Catalysts that speed up reactions are important in industry to reduce costs.

Catalysts change the rates of chemical reactions. Most catalysts are used to speed up reactions. The catalyst is left at the end of the reaction and so it can be used over and over again. Catalysts work by lowering the activation energy of a reaction so that more collisions result in a reaction.

Although some catalysts are expensive, they can be economical because they do not need replacing very often. They can also reduce the energy costs and time needed for a reaction. Catalysts often work with only one type of reaction and so different reactions need different catalysts. Finding new and better catalysts is a major area of research for the chemical industry.

Key words: catalyst, activation energy

AQA EXAMINER SAYS...

Most of the catalysts that are used are positive catalysts that increase the rates of reactions. (Negative catalysts or inhibitors are used to slow reactions down, but are not studied at GCSE.)

GET IT RIGHT!

Catalysts are left at the end of a reaction but they definitely play a part in the reaction.

CHECK YOURSELF

1 What is the effect of a catalyst on the activation energy of a reaction?
2 Why can expensive catalysts be economical?
3 Why are many different catalysts needed?

C2 4 End of chapter questions

1 **Write an equation to show how to calculate the rate of a reaction.**

2 **Suggest one way that you could measure the rate of reaction between magnesium and hydrochloric acid:** $Mg + 2HCl \longrightarrow MgCl_2 + H_2$

3 **What name is used for the minimum energy needed for a reaction to happen?**

4 **How could you increase the rate of the reaction between magnesium and hydrochloric acid?**

5 **Why do many reactions go twice as fast if the temperature is increased by 10°C?**

6 **What will happen to the rate of a reaction if the temperature is decreased from 30°C to 10°C?**

7 **Explain why the frequency of collisions increases if you increase the concentration of a solution.**

8 **Why do gases react faster at higher pressures?**

9 **How do catalysts speed up reactions?**

10 **Why do catalysts not need replacing very often?**

C2 5 Pre Test: Energy and reactions

1. **What name do we use for reactions that transfer energy to the surroundings?**
2. **What type of reaction causes the temperature of the surroundings to decrease?**
3. **In a reversible reaction, how does the energy of the forward reaction compare with the energy of the reverse reaction?**
4. **How does increasing the temperature affect a reversible reaction in a closed system? [Higher Tier only]**
5. **What sort of reversible reactions are affected by pressure? [Higher Tier only]**
6. **Why are compromise conditions used in the process to make ammonia? [Higher Tier only]**

students' book page 148

C2 5.1 Exothermic and endothermic reactions

KEY POINTS

1. Exothermic reactions transfer energy to the surroundings.
2. Endothermic reactions transfer energy from the surroundings.

When chemical reactions take place energy is transferred as bonds are broken and made. Reactions that transfer energy to the surroundings are called 'exothermic' reactions. The energy transferred often heats up the surroundings and so the temperature increases. Exothermic reactions include combustion, such as burning fuels and metals, respiration and neutralisation.

Endothermic reactions take in energy from the surroundings. Some cause a decrease in temperature and others require a supply of energy. When some solid compounds are mixed with water, the temperature decreases because endothermic reactions happen. Thermal decomposition reactions need a supply of heat to keep going. Photosynthesis is an important endothermic reaction that uses light energy.

When a fuel burns in oxygen, energy is transferred to the surroundings

AQA EXAMINER SAYS...

You should know the main types of reaction that are exothermic and endothermic reactions. You should be able to recognise a reaction as exothermic or endothermic from information about temperature changes.

EXAM HINTS

Endothermic reations take in heat. Heat exits (transfers from) exothermic reactions.

When we eat sherbet we can feel an endothermic reaction! Sherbet dissolving in the water in your mouth takes in energy – giving a slight cooling effect.

Investigating energy changes

The thermometer is used to measure the temperature change which takes place during the reaction.

Chemicals are mixed in the cup. The insulation reduces the rate at which energy can enter or leave the contents of the cup.

Styrofoam cup

We can use very simple apparatus to investigate the energy changes in reactions. Often we don't need to use anything more complicated than a Styrofoam drinks cup and a thermometer.

CHECK YOURSELF

1 What do we call reactions that transfer energy from the surroundings?

2 How do we know that burning methane is exothermic?

3 Why do some sweets produce a cooling effect in your mouth?

students' book page 150

C2 5.2 Energy and reversible reactions

KEY POINTS

1 Reversible reactions are exothermic in one direction and endothermic in the other direction.
2 Increasing the temperature increases the amount of products from the endothermic reaction. [Higher Tier only]

GET IT RIGHT!

The energy changes of the forward and reverse reactions are always equal.

In reversible reactions, the forward and reverse reactions involve equal but opposite energy transfers. A reversible reaction that is exothermic in one direction must be endothermic in the other direction. The amount of energy released by the exothermic reaction exactly equals the amount taken in by the endothermic reaction.

Changing the temperature of a reversible reaction in a closed system changes the amounts of the reactants and products. If we increase the temperature, the amount of products from the endothermic reaction increases. If we decrease the temperature, the amount of products from the exothermic reaction increases. This means that we can change the yield of the reaction by changing the temperature.

- Heating blue copper sulfate crystals is an endothermic reaction:

blue crystals		white powder	
$CuSO_4.5H_2O$	$\rightleftharpoons$	$CuSO_4$	$+ 5H_2O$
hydrated copper sulfate		anhydrous copper sulfate	

- Adding water to anhydrous copper sulfate is an exothermic reaction.

students' book page 150

C2 5.2 Energy and reversible reactions (cont.)

AQA EXAMINER SAYS...

You will be given information about which reaction is exothermic or endothermic in questions about reversible reactions. Remember that equations for reversible reactions should balance, just like all chemical equations.

HIGHER

In a closed system, the relative amounts of reactants and products in a reversible reaction depend on the temperature. This is a very important consideration in many industrial processes.

Look at the table below:

If a reaction is exothermic	If a reaction is endothermic ......
... an increase in temperature decreases the yield of the reaction, so the amount of products formed is lower.	... an increase in temperature increases the yield of the reaction, so the amount of products formed is larger.
... a decrease in temperature increases the yield of the reaction, so the amount of products formed is larger.	... a decrease in temperature decreases the yield of the reaction, so the amount of products formed is lower.

CHECK YOURSELF

1. In a reversible reaction the forward reaction is endothermic. What does this tell you about the reverse reaction?
2. In a reversible reaction the forward reaction releases 50 kJ of energy. What will be the energy transfer for the reverse reaction?
3. In the reversible reaction $H_2 + I_2 \rightleftharpoons 2HI$ the formation of HI is exothermic. What should you do to the temperature to increase the yield of HI? [Higher Tier only]

students' book page 152

C2 5.3 More about the Haber process

KEY POINTS

1. For reversible reactions involving gases, increasing the pressure increases the yield of the reaction that produces the smaller number of molecules of gas.
2. The conditions for industrial processes are chosen to give as much product as possible as quickly as possible for the lowest cost.

HIGHER

Changes in pressure affect the yield of reversible reactions that have different numbers of molecules of gases in the reactants and products. An increase in pressure will increase the yield of a reaction that has fewer molecules of gases in the products than in the reactants.

If a reaction produces a larger volume of gases	If a reaction produces a smaller volume of gases ......
... an increase in pressure decreases the yield of the reaction, so the amount of products formed is lower.	... an increase in pressure increases the yield of the reaction, so the amount of products formed is larger.
... a decrease in pressure increases the yield of the reaction, so the amount of products formed is larger.	... a decrease in pressure decreases the yield of the reaction, so the amount of products formed is lower.

In the reaction for the Haber process:

$$N_2 + 3H_2 \rightleftharpoons 2NH_3$$

four molecules of reactant gases produce two molecules of ammonia gas. So increasing the pressure will produce more ammonia. However, increasing the pressure increases the costs of the process and so a compromise of a reasonably high pressure is used.

EXAMINER SAYS...

Use the balanced equation to count the total number of molecules of gases in the reactants and compare this with the total number of molecules of gases in the products. Count only the gases, ignoring any solids and liquids in the equation.

GET IT RIGHT!

Changing the pressure only affects the yield in reactions with different numbers of molecules of gases in the reactants compared with the products.

HIGHER

The reaction to produce ammonia is exothermic, so lower temperatures give higher yields. However, the reaction is slower at lower temperatures because the rate decreases and the catalyst does not work as well, so a compromise temperature is used. The conditions are chosen to produce ammonia as economically as possible. Industrial processes are being developed that use low temperatures and pressures to reduce energy use and waste.

It is expensive to build chemical plants that operate at high pressures

CHECK YOURSELF

1 How can you tell if a reversible reaction will be affected by changes in pressure?

2 Why do many industrial processes making ammonia operate at about 200 atmospheres pressure?

3 Why is a temperature of about 450°C used to make ammonia?

C2 5 End of chapter questions

1 **Name two types of reaction that are exothermic.**

2 **Why do some solids produce a cooling effect when mixed with water?**

3 **The reaction $2SO_2 + O_2 \rightleftharpoons 2SO_3$ is exothermic in the forward direction. What should be done to the temperature to obtain a high yield of SO_3? [Higher Tier only]**

4 **In the forward reaction in question 3, there are 95 kJ of energy released. What is the energy change for the reverse reaction?**

5 **The reaction $H_2 + I_2 \rightleftharpoons 2HI$ is reversible. How would increasing the pressure affect the yield of HI? [Higher Tier only]**

6 **The reaction $2NO_2 \rightleftharpoons N_2O_4$ is exothermic. What conditions of temperature and pressure would increase the yield of N_2O_4? [Higher Tier only]**

C2 6 Pre Test: Electrolysis

1. **What happens to ionic compounds when they are electrolysed?**
2. **What substances can be produced at the positive electrode in electrolysis?**
3. **Where does reduction take place during electrolysis?**
4. **What sort of electrolyte can produce hydrogen at the negative electrode?**
5. **What products are formed when brine is electrolysed?**
6. **Why is electrolysis of brine done on a large scale in industry?**
7. **What change takes place at the positive electrode in the purification of copper by electrolysis?**
8. **What happens to the impurities from the copper?**

students' book page 158

C2 6.1 Electrolysis – the basics

KEY POINTS

1. Electrolysis decomposes ionic substances into elements.
2. Ions are free to move when ionic solids are molten or dissolved in water.
3. Metals or hydrogen are formed at the negative electrode and non-metallic elements are formed at the positive electrode.

AQA EXAMINER SAYS...

Electrodes are usually made of substances that will not react with the electrolyte or the elements that are formed.

When electricity is passed through a molten ionic compound or a solution containing ions, electrolysis takes place. The molten ionic solid or solution of ions is called the 'electrolyte'.

The electrical circuit has two conducting rods called 'electrodes' that make contact with the electrolyte. The ions in the electrolyte move to the electrodes where they are discharged to produce elements.

- Positively charged ions are attracted to the negative electrode where they form metals. Hydrogen may be formed at the negative electrode if the ions are dissolved in water.
- Negatively charged ions are attracted to the positive electrode where they lose their charge to form non-metallic elements.
- For example, when molten lead bromide is electrolysed, lead and bromine are produced.

Key words: electrolysis, electrolyte, electrode

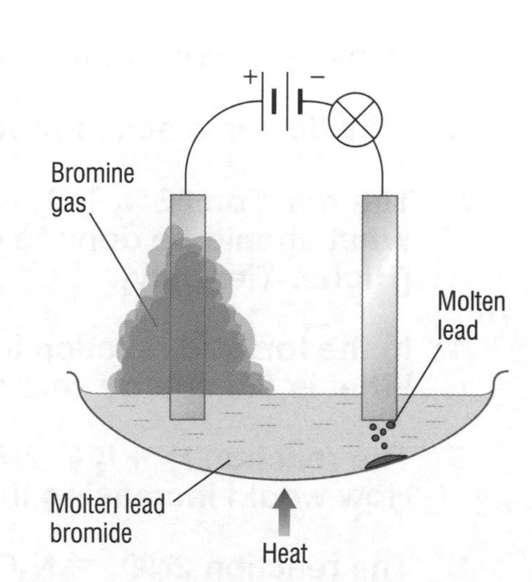

When we pass electricity through molten lead bromide it forms molten lead and brown bromine gas, as the electrolyte is broken down by the electricity

GET IT RIGHT!

Ions are discharged at the electrodes to produce elements.

CHECK YOURSELF

1 What is an electrolyte?

2 What happens to positively charged ions during electrolysis?

3 Which elements are produced when molten zinc chloride is electrolysed?

students' book page 160

C2 6.2 Changes at the electrodes

KEY POINTS

1 Positively charged ions gain electrons at the negative electrode.
2 Negatively charged ions lose electrons at the positive electrode.
3 Gaining electrons is reduction; loss of electrons is oxidation.
4 Electrolysing solutions of ions in water may produce hydrogen.

- When positively charged ions reach the negative electrode they gain electrons to become neutral atoms. Gaining electrons is called 'reduction', so the positive ions have been reduced. Ions with a single positive charge gain one electron and those with a 2+ charge gain 2 electrons.
- At the positive electrode, negative ions lose electrons to become neutral atoms. This is 'oxidation'. Some non-metal atoms combine to form molecules, for example bromine forms Br_2.

HIGHER

We can represent the changes at the electrodes by half equations.
The equations for lead bromide are:

At the negative electrode: $Pb^{2+} + 2e^- \rightarrow Pb$

At the positive electrode: $2Br^- \rightarrow Br_2 + 2e^-$

Only Higher Tier students will be expected to write half equations for reactions at electrodes in the examination.

Water contains hydrogen ions and hydroxide ions. When solutions of ions in water are electrolysed, hydrogen may be produced at the positive electrode. This happens if the other positive ions are of metals more reactive than hydrogen.

Key words: reduction, oxidation

EXAM HINTS

In half equations the number of electrons must balance the number of charges on the ion.

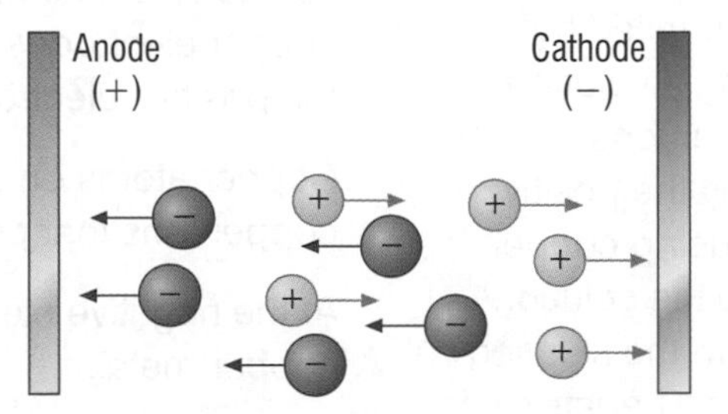

An ion always moves towards the oppositely charged electrode

GET IT RIGHT!

Oxidation Is Loss of electrons, Reduction Is Gain – OIL RIG.

CHECK YOURSELF

1 What happens in electrolysis to copper ions, Cu^{2+}, at the negative electrode?

2 What happens in electrolysis to chloride ions, Cl^-, at the positive electrode?

3 Why are ions of metals always reduced in electrolysis?

students' book page 162

C2 6.3 Electrolysing brine

KEY POINTS

1. Electrolysing brine produces hydrogen, chlorine and sodium hydroxide.
2. The products have many important uses.

You may be asked to apply your knowledge of electrolysis to other industrial processes in the examination.

Brine is a solution of sodium chloride in water. When it is electrolysed, hydrogen is produced at the negative electrode from hydrogen ions in the water. Chlorine is produced at the positive electrode from the chloride ions. This leaves sodium ions and hydroxide ions (from water) in the solution.

HIGHER

The half equations for the reactions at the electrodes are:

At the positive electrode: $2Cl^- \rightarrow Cl_2 + 2e^-$

At the negative electrode: $2H^+ + 2e^- \rightarrow H_2$

- Sodium hydroxide is a strong alkali and has many uses including making soap, making paper, making bleach, neutralising acids and controlling pH.
- Chlorine is used to kill bacteria in drinking water and swimming pools, and to make bleach, disinfectants and plastics.
- Hydrogen is used to make margarine and hydrochloric acid.

Key words: brine, hydrogen, chlorine, sodium hydroxide

CHECK YOURSELF

1. What are the four ions present in brine?
2. Explain why sodium hydroxide is left in the solution.
3. Give two uses for each of the products.

students' book page 164

C2 6.4 Purifying copper

KEY POINTS

1. Copper can be purified by electrolysis using a solution containing copper ions.
2. Impure copper is the positive electrode, producing copper ions that go into the solution.
3. Copper ions from the solution are reduced at the negative electrode to form pure copper.

Impurities in copper affect its properties including its conductivity. Copper for use as electrical wires must be very pure. It can be purified by electrolysis, using copper electrodes in a solution of a copper salt. The impure copper is used as the positive electrode and the negative electrode is a thin sheet of pure copper.

Copper atoms on the positive electrode are oxidised, losing electrons to form copper ions that go into the solution.

At the negative electrode copper ions from the solution are reduced, forming copper metal.

The copper is deposited on the negative electrode, which increases in thickness.

HIGHER

The half equations for the reactions at the electrodes are:

At the positive electrode: $Cu(s) \rightarrow Cu^{2+}(aq) + 2e^-$

At the negative electrode: $Cu^{2+}(aq) + 2e^- \rightarrow Cu(s)$

As the copper from the positive electrode dissolves, the impurities are released and collect as sludge at the bottom of the cell. The impurities include precious metals like gold, silver and platinum. These are extracted from the sludge.

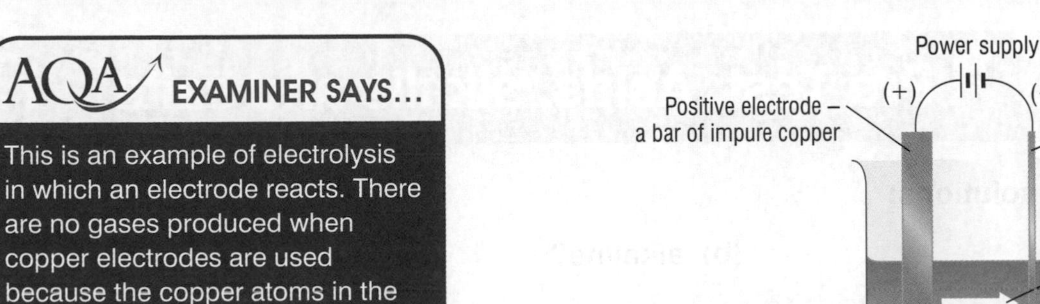

Copper is refined using electrolysis

CHECK YOURSELF

1 Why is it necessary for copper wires to be very pure?

2 Explain why copper ions from the positive electrode go into the solution when copper is purified by electrolysis.

3 Why does the negative electrode increase in thickness?

C2 6 End of chapter questions

1 **What happens to the ions in an electrolyte during electrolysis?**

2 **What is produced at the positive electrode when molten magnesium chloride is electrolysed?**

3 **Explain what happens to magnesium ions at the negative electrode when magnesium chloride is electrolysed.**

4 **What product is formed by oxidation when copper bromide is electrolysed?**

5 **Why does electrolysing brine give different products to electrolysing molten sodium chloride?**

6 **Which products from electrolysing brine are used to make bleach?**

7 **What happens to copper atoms from the impure copper electrode in the purification of copper by electrolysis?**

8 **Why is pure copper deposited on the other electrode?**

9 **Copy and complete the balanced half equation for the deposition of aluminium at the negative electrode:**

$Al^{3+} \rightarrow Al$ **[Higher Tier only]**

10 **Copy and complete the balanced half equation that shows the release of chlorine gas at the positive electrode:**

$Cl^{-} \rightarrow Cl_2$ **[Higher Tier only]**

C2 7 Pre Test: Acids, alkalis and salts

1. Which ions make solutions:
   (a) acidic
   (b) alkaline?
2. A solution has a pH of 12. What does this tell you about the solution?
3. What are the products when an acid reacts with a metal?
4. Name a base that could be used to make magnesium chloride.
5. What are the products formed when potassium hydroxide reacts with sulfuric acid?
6. How can you make the insoluble salt lead chloride?

students' book page 170

C2 7.1 Acids and alkalis

KEY POINTS

1. When acids are added to water they form hydrogen ions, $H^+(aq)$.
2. When alkalis are added to water they form hydroxide ions, $OH^-(aq)$.
3. The pH scale measures the acidity or alkalinity of solutions in water.

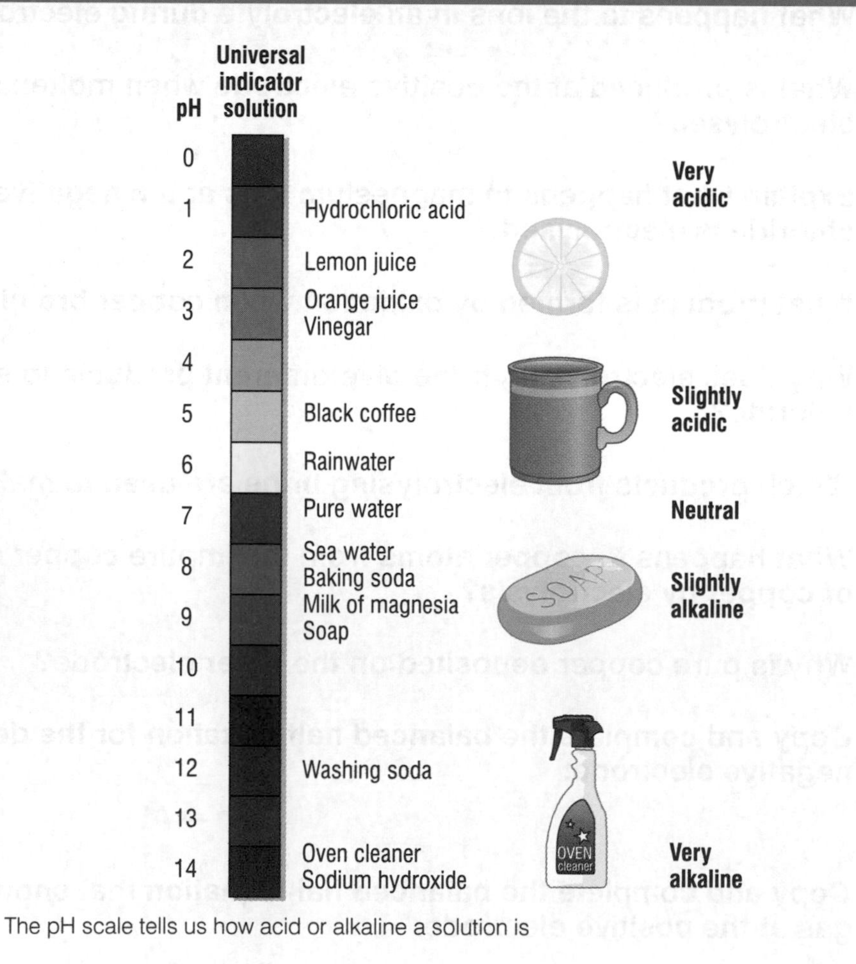

The pH scale tells us how acid or alkaline a solution is

AQA EXAMINER SAYS...

A base is a substance that will neutralise an acid, but some bases do not dissolve in water. Only bases that dissolve in water are called alkalis.

GET IT RIGHT!

Acids have pH values below 7. Alkalis have pH values greater than 7.

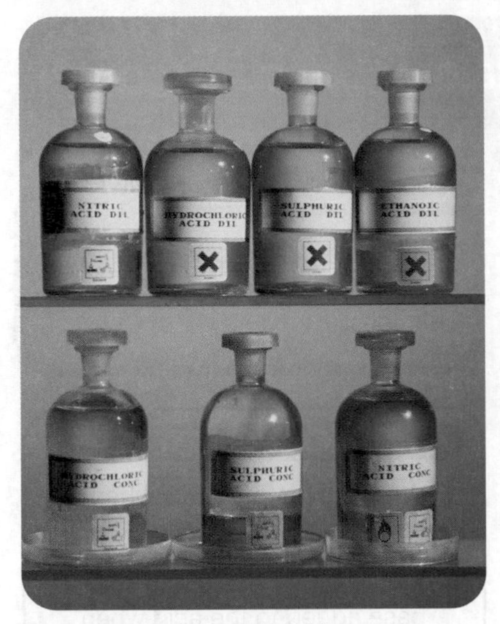

Some common laboratory acids

- Pure water is neutral and has a pH value of 7.
- Acids are substances that produce hydrogen ions, $H^+(aq)$, when they are added to water. This makes the solution acidic and it has a pH value of less than 7.
- Bases react with acids and neutralise them. Alkalis are bases that dissolve in water to make the solution alkaline. They produce hydroxide ions, $OH^-(aq)$, in the solution. Alkaline solutions have a pH value greater than 7.

The pH scale has values from 0 to 14. Solutions that are very acidic have low pH values between 0 and 2, and solutions that are very alkaline have high pH values of 12 to 14. Indicators have different colours in acidic and alkaline solutions. Universal indicator and full-range indicators have different colours at different pH values.

Key words: hydrogen ions, hydroxide ions, pH, neutral, neutralise, indicators

CHECK YOURSELF

1 When a solid dissolves in water the solution has a pH value of 3.5. Name the ions that give this pH value.

2 Why does the pH of water change when alkalis dissolve?

3 Name a substance that can be used to test the pH of a solution.

students' book page 172

C2 7.2 Making salts from metals or bases

KEY POINTS

1 Salts are formed when acids react with metals or bases.
2 All of the acid can be used up if we add excess solid.
3 The salt that is produced depends on the acid and the metal in the reactants.

We can make salts by reacting acids with metals or bases. Acids will react with metals that are above hydrogen in the reactivity series. These metals react with acids to form hydrogen gas and a salt.

$$\text{ACID} + \text{METAL} \rightarrow \text{SALT} + \text{HYDROGEN}$$

However, the reactions with the most reactive metals are too violent to be done safely.

Bases are metal oxides or metal hydroxides. They react with acids to form a salt and water.

$$\text{ACID} + \text{BASE} \rightarrow \text{SALT} + \text{WATER}$$

A metal, or a base that is insoluble in water, is added a little at a time to the acid until all of the acid has reacted. The mixture is then filtered to remove the excess solid, leaving a solution of the salt. The solid salt is made when water is evaporated from the solution so that it crystallises.

Chlorides are made from hydrochloric acid, nitrates from nitric acid and sulfates from sulfuric acid.

Key words: excess, chlorides, nitrates, sulfates

AQA EXAMINER SAYS...

You should find it easy to work out the name of the salt from the names of the acid and the metal or base. Working out which acid and base or metal you need to make a named salt can be a bit trickier. It is usually safest to use metal oxides.

C2 7.2 Making salts from metals or bases (cont.)

Making crystals of copper sulfate

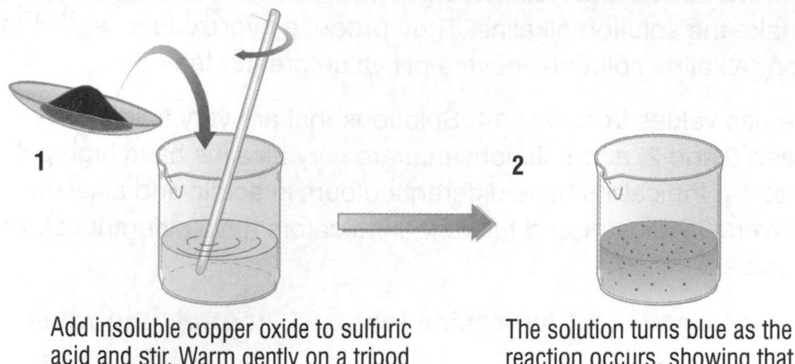

Add insoluble copper oxide to sulfuric acid and stir. Warm gently on a tripod and gauze (do not boil).

The solution turns blue as the reaction occurs, showing that copper sulfate is being formed

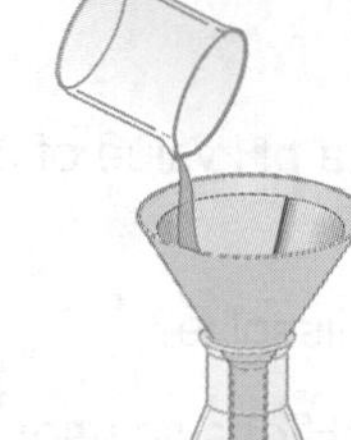

When the reaction is complete, filter the solution to remove excess copper oxide

We can evaporate the water so that crystals of copper sulfate are left

GET IT RIGHT!

Watch the spelling of hydrochloric acid, and remember that it forms chlorides.

CHECK YOURSELF

1. What are the products when an acid reacts with a base?
2. Why is an excess of the solid base added to the acid when making a salt?
3. Name the salt made from hydrochloric acid and zinc oxide.
4. Name the acid and base you would use to make copper nitrate.

C2 7.3 Making salts from solutions

KEY POINTS

1. An indicator is used to find when an acid has exactly reacted with an alkali to form a salt.
2. We can make an insoluble salt by mixing two solutions containing the ions in the salt.
3. Ions can be removed from solutions by precipitation.

We can make soluble salts by reacting an acid and an alkali.

ACID + ALKALI → SALT + WATER

We can summarise the reaction between any acid and alkali by just showing the ions that react:

$$H^+ (aq) + OH^- (aq) \rightarrow H_2O (l)$$

(from acid) (from alkali)

However, there is no visible change when the solutions react so we need an indicator to show when the reaction is complete. The indicator can be removed from the solution after the reaction.

Alternatively, the volumes of acid and alkali needed to produce the salt are found and these volumes of fresh solutions are mixed. The pure salt is obtained by crystallisation.

Ammonia solution is an alkali that does not contain a metal. It forms ammonium salts, such as ammonium nitrate, which are used as fertilisers.

We can make insoluble salts by mixing solutions of soluble salts that contain the ions needed. For example, we can make lead iodide by mixing solutions of lead nitrate and potassium iodide. The lead iodide forms a precipitate that can be filtered and dried.

Some pollutants can be removed from water as precipitates by adding ions that react with them to form insoluble salts.

Key words: ammonia, ammonium, precipitate

Making lead chloride

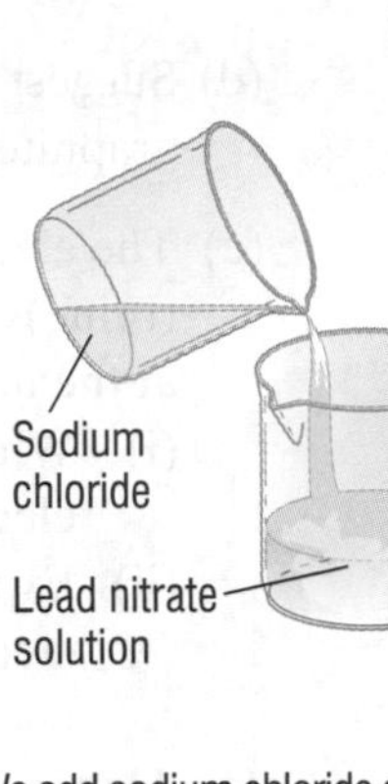

1 We add sodium chloride solution to lead nitrate solution and stir

2 The precipitate of lead chloride that forms is filtered off from the solution

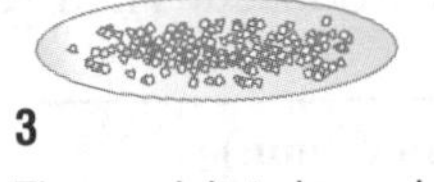

3 The precipitate is washed with distilled water and dried

You do not need to remember which salts are soluble or insoluble in water because you will be told about their solubility in any questions in the examination.

CHECK YOURSELF

1 Sodium sulfate can be made by reacting sodium hydroxide and sulfuric acid. What other substance will you need to add to the reaction mixture?

2 Name the salt formed when ammonia solution is reacted with hydrochloric acid.

3 Complete the equation for the reaction to make insoluble copper carbonate:

… sulfate + sodium … → copper carbonate + sodium sulfate

C2 7 End of chapter questions

1 **A solution has a pH value of 2. What does this tell you about the solution?**

2 **What is an alkali?**

3 **Name the products of the reaction between magnesium and sulfuric acid.**

4 **Name the products of the reaction between hydrochloric acid and copper oxide.**

5 **Why is an indicator needed when reacting an acid with an alkali?**

6 **Name the products when sodium hydroxide solution reacts with nitric acid.**

7 **Write an equation to show how the ions in an acid and alkali react so that it becomes neutral.**

8 **Suggest the name of a solution you could add to lead nitrate to produce insoluble lead sulfate.**

1 Hydrogen peroxide decomposes to produce oxygen gas and water:

$$2H_2O_2(aq) \rightarrow 2H_2O(aq) + O_2(g)$$

The reaction is catalysed by manganese(IV) oxide.

Some students added 2 g of manganese(IV) oxide to 20 cm^3 of hydrogen peroxide solution and measured the volume of gas produced. Their results are shown in the table.

Time (min)	0	1	2	3	4	5	6	7	8	9
Volume of gas (cm^3)	0	18	34	48	59	65	74	78	80	80

(a) Plot a graph of the results with time on the horizontal axis and volume of gas on the vertical axis. Draw a smooth line through the points, omitting any result that is anomalous. (4 marks)

(b) (i) How can you tell from the graph that the rate of reaction was fastest at the start of this experiment? (1 mark)

(ii) Explain, in terms of particles, why the rate of reaction was fastest at the start. (2 marks)

(c) The students repeated the experiment with 2 g of manganese(IV) oxide that was more finely powdered. All other conditions were kept the same. Sketch a line on the same axes to show the results you would expect for this experiment. (2 marks)

2 The electrolysis of molten sodium chloride is used to produce sodium metal. The diagram shows the type of electrolysis cell that is used.

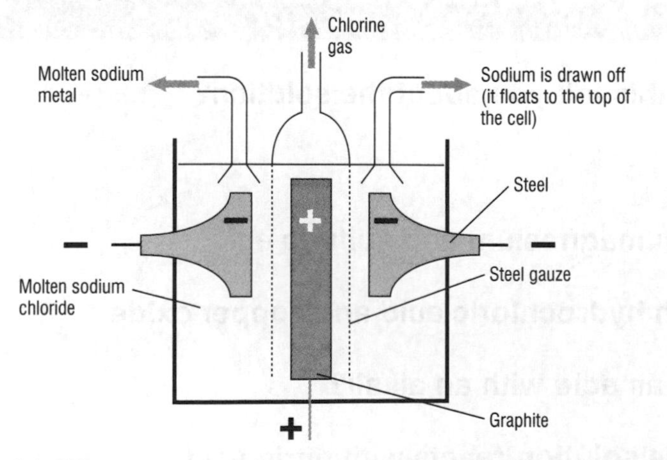

(a) Why must the sodium chloride be molten? (1 mark)

(b) Explain how sodium is produced at the negative electrode. (2 marks)

(c) Explain how chlorine is produced at the positive electrode. (2 marks)

(d) Suggest why the positive electrode is made of graphite. (1 mark)

(e) The electrolysis of aqueous sodium chloride solution (brine) is also done in industry. Hydrogen is produced at the negative electrode.

(i) What are the other two products when brine is electrolysed? (2 marks)

(ii) Explain why hydrogen is produced at the negative electrode. (2 marks)

3 The Haber Process is used to make ammonia.

(a) The table shows the percentage yield of ammonia at different temperatures and pressures.

Pressure (atmospheres)	Percentage yield of ammonia at 350°C (%)	Percentage yield of ammonia at 500°C (%)
50	25	5
100	37	9
200	52	15
300	63	20
400	70	23
500	74	25

(i) Draw graphs of this data on the same axes. Put the percentage yield of ammonia (%) on the vertical axis and pressure (atmospheres) on the horizontal axis. Plot the points and draw a smooth line for each temperature. Label each line with its temperature. (4 marks)

(ii) Use your graphs to find the conditions needed to give a yield of 30% ammonia. (1 mark)

(iii) On the same axes, sketch the graph you would expect for a temperature of 450°C. (1 mark)

(b) This equation represents the reaction in which ammonia is formed:

$$N_2(g) + 3H_2(g) \rightleftharpoons 2NH_3(g)$$

(i) What does the symbol $\rightleftharpoons$ in this equation tell you about the reaction? (1 mark)

(ii) Explain why a temperature of 450°C is used in industrial processes to make ammonia. (2 marks)

(iii) Explain why a pressure of 200 atmospheres is used in industrial processes to make ammonia. (2 marks)

[Higher]

The answer is worth 4 out of the 5 marks available.

The responses worth a mark are underlined in red.

We can improve the answer in several ways:

Explain as fully as you can why hydrogen is produced at the negative electrode when a solution of sodium chloride in water is electrolysed. *(5 marks)*

Sodium chloride solution contains Na^+ ions and H^+ ions. The sodium ions come from sodium chloride and the hydrogen ions from water. Both these positive ions are attracted to the negative electrode but only the hydrogen ions are reduced because they are less reactive than sodium ions. The reaction that takes place at the electrode is $H^+ + e^- \rightarrow H$.

The answer is clearly expressed and begins well. However, the reason that hydrogen ions are reduced in preference to sodium ions is incorrect. It is sodium metal that is more reactive than hydrogen, and therefore sodium ions are more difficult to reduce.

It is a good idea to include equations for reactions when possible, but the reaction here is incomplete and shows only the formation of hydrogen atoms, not hydrogen gas, which is made of molecules of H_2. However, if reduction had not been mentioned it could have gained the mark for reducing hydrogen ions.

The answer is worth 4 marks out of the 7 marks available.

The responses worth a mark are underlined in red.

We can improve the answer in several ways:

Antacid tablets contain calcium carbonate. They neutralise excess acid in the stomach.

(a) How does the pH in the stomach change when the tablets 'neutralise excess acid'? *(1 mark)*

(b) Write an equation to show the reaction between the ions in the neutralisation reaction. *(2 marks)*

(c) Chewing the tablet cures indigestion faster than swallowing the tablet in one piece. Explain why, as fully as you can, using particle theory. *(4 marks)*

(a) The pH gets less.

(b) $H^+ + OH^- \rightarrow H_2O$

(c) Chewing the tablet breaks it down into smaller pieces. The smaller pieces have more surface area, so the acid can get to them more easily. Particles of acid bump into the bits of tablet and when they do this they react. Because the bits of tablet are smaller the indigestion is cured faster.

This is incorrect and scores no mark. The student has confused acidity with pH. The acidity decreases ('gets less'), but the pH increases when the acid is neutralised.

The answer starts well, although 'get to them more easily' is not the same as 'a faster reaction'.

The increase in frequency of collisions and the increase in rate of reaction are not made clear.

This sentence just repeats information from earlier in the answer. Replace with 'Particles of acid bump into the bits of tablet more frequently and so the rate of the reaction increases.'

P2 | Additional physics (Chapters 1–3)

Checklist

This spider diagram shows the topics in the first half of the unit. You can copy it out and add your notes and questions around it, or cross off each section when you feel confident you know it for your exams.

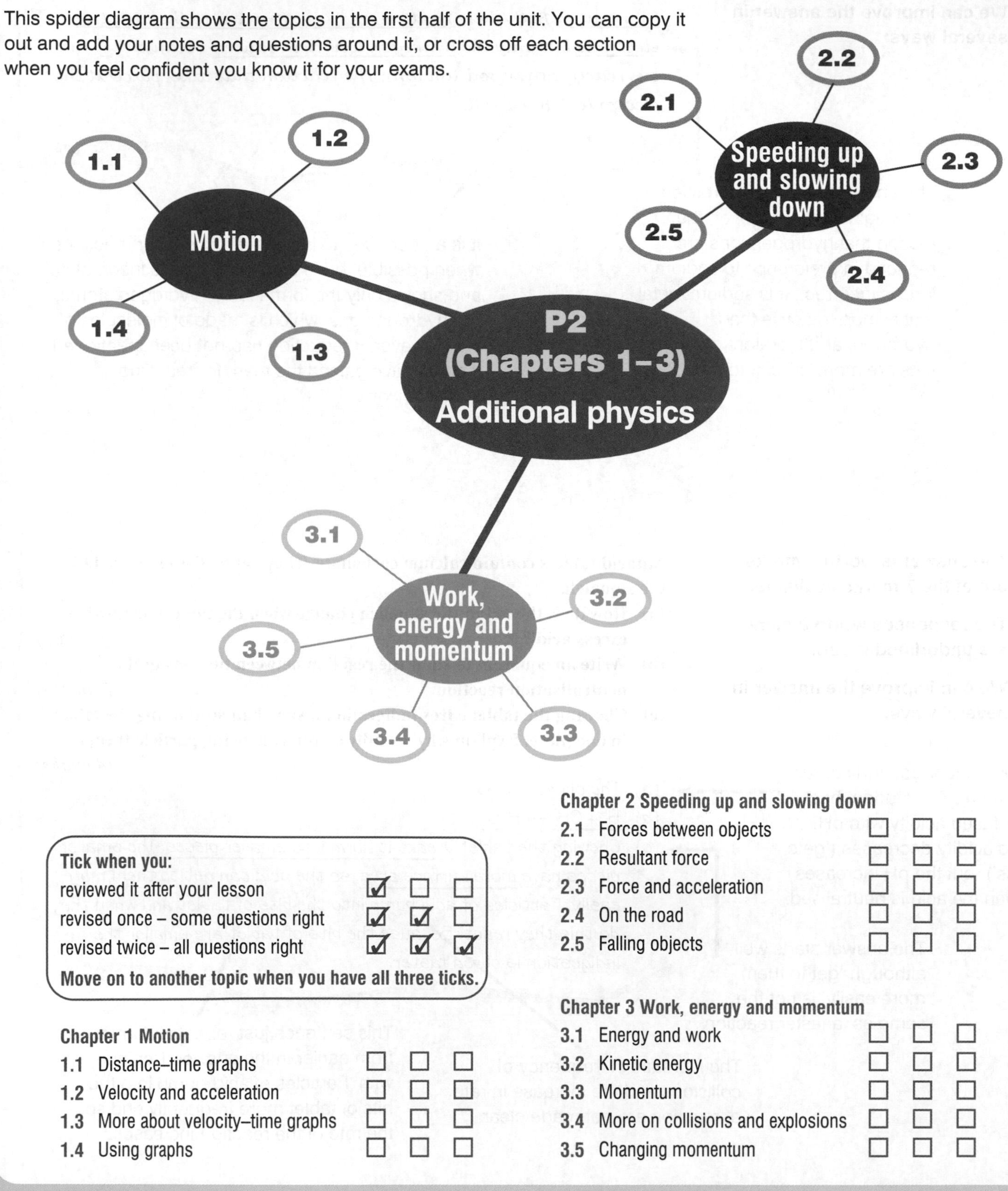

Tick when you:

reviewed it after your lesson	✓	☐	☐
revised once – some questions right	✓	✓	☐
revised twice – all questions right	✓	✓	✓

Move on to another topic when you have all three ticks.

Chapter 1 Motion

1.1 Distance–time graphs	☐	☐	☐
1.2 Velocity and acceleration	☐	☐	☐
1.3 More about velocity–time graphs	☐	☐	☐
1.4 Using graphs	☐	☐	☐

Chapter 2 Speeding up and slowing down

2.1 Forces between objects	☐	☐	☐
2.2 Resultant force	☐	☐	☐
2.3 Force and acceleration	☐	☐	☐
2.4 On the road	☐	☐	☐
2.5 Falling objects	☐	☐	☐

Chapter 3 Work, energy and momentum

3.1 Energy and work	☐	☐	☐
3.2 Kinetic energy	☐	☐	☐
3.3 Momentum	☐	☐	☐
3.4 More on collisions and explosions	☐	☐	☐
3.5 Changing momentum	☐	☐	☐

What are you expected to know?

Chapter 1 Motion (See students' book pages 184–195)

- Construction and use of distance–time graphs.
- Construction and use of velocity–time graphs.
- Acceleration = change in speed/time taken for the change

Chapter 2 Speeding up and slowing down (See students' book pages 196–209)

- Weight = mass × gravitational field strength
- What is meant by a resultant force?
- Resultant forces produce accelerations.
- Resultant force = mass × acceleration
- Factors that affect stopping distance.
- Factors that affect reaction time and braking distance.
- Frictional forces oppose motion.
- Bodies falling through fluids reach a terminal velocity.

Chapter 3 Work, energy and momentum (See students' book pages 210–223)

- Work done = energy transferred
- Work done = force × distance moved in the direction of the force
- Kinetic energy = $\frac{1}{2}$ mass × speed^2
- Momentum = mass × velocity
- Total momentum is conserved in a collision or explosion provided no external forces act.
- Force = change in momentum/time taken for the change

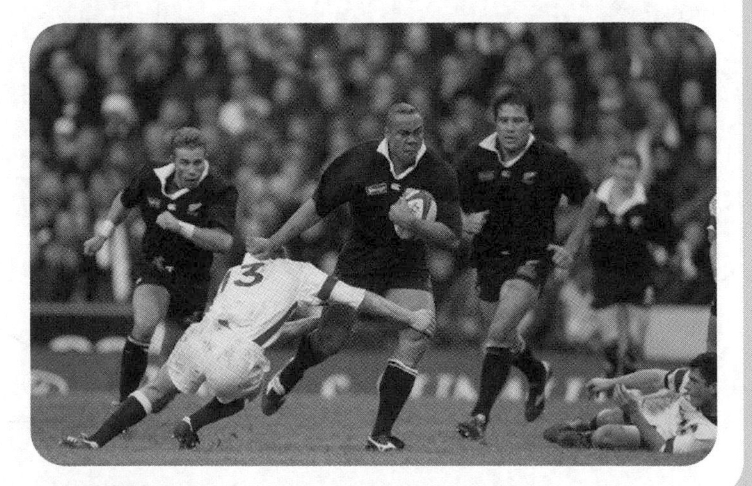

P2 1 Pre Test: Motion

1. What is the SI unit of speed?
2. What does the distance–time graph for a stationary body look like?
3. What is the equation that allows us to calculate acceleration?
4. What does a negative value for acceleration mean?
5. What does the slope of the line on a velocity–time graph represent?
6. What would the velocity–time graph for a body moving at a constant speed look like?
7. How do you calculate the slope of a line on a graph? [Higher Tier only]
8. What happens to the slope of the line on a distance–time graph if the speed increases?

students' book page 184

P2 1.1 Distance–time graphs

KEY POINTS

1. The steeper the line on a distance–time graph, the greater the speed it represents.
2. Speed (metre/second, m/s) = $\frac{\text{distance travelled (metre, m)}}{\text{time taken (second, s)}}$

Always show the stages in your working when you do calculations. Always include a unit with your answer if one is not given.

We can use graphs to help us describe the motion of a body.

A distance–time graph shows the distance of a body from a starting point (*y*-axis) against time taken (*x*-axis).

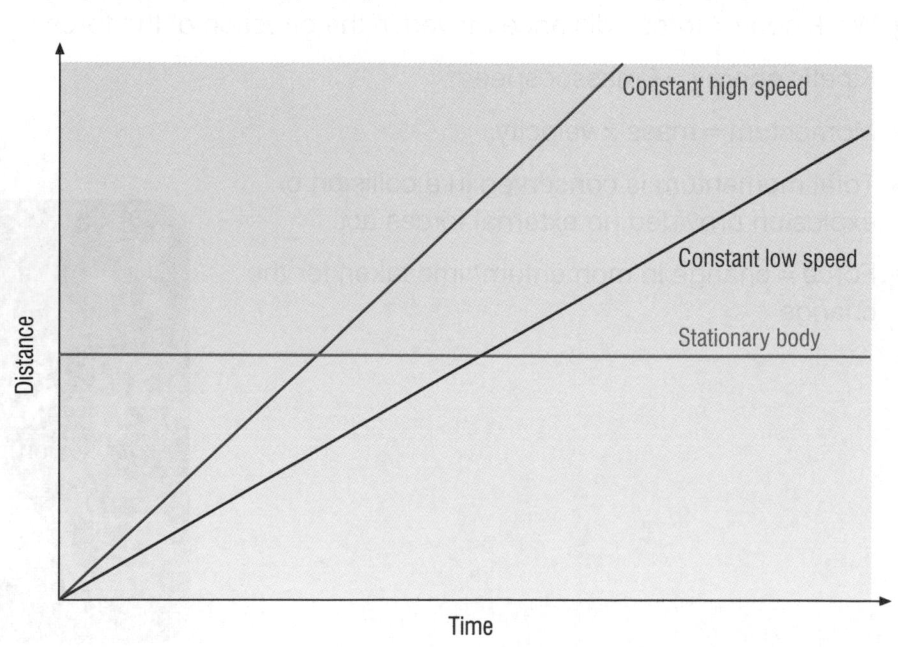

Distance–time graphs

GET IT RIGHT!

Make sure that you always label the axes on a graph with a quantity and a unit.

The slope of the line on a distance–time graph represents speed.

The steeper the slope, the greater the speed.

The speed of a body is the distance travelled each second. We can calculate the speed of a body using the equation:

$$\text{speed} = \frac{\text{distance travelled}}{\text{time taken}}$$

The SI unit of speed is metres per second (m/s).

Key words: graph, distance, time, metre, second, slope

CHECK YOURSELF

1. What does the slope of the line on a distance–time graph represent?
2. What is the equation that relates speed, distance and time?
3. What is the SI unit of distance?

students' book page 186

P2 1.2 Velocity and acceleration

KEY POINTS

1. Velocity is speed in a given direction.
2. Acceleration is change of velocity per second.
3. A body travelling at a steady speed is accelerating if its direction is changing.

The velocity of a body is its speed in a given direction. If the body changes direction it changes velocity, even if its speed stays the same.

If the velocity of a body changes, we say that it accelerates.

We can calculate acceleration using the equation:

$$\text{acceleration} = \frac{\text{change in velocity}}{\text{time taken for the change}}$$

The SI unit of acceleration is metres per second squared (m/s^2).

If the value calculated for acceleration is negative, the body is decelerating – slowing down.

Key words: velocity, direction, acceleration, decelerating

GET IT RIGHT!

Remember that a velocity must include a direction.

If a body changes direction, it changes velocity. So it accelerates, even if its speed stays the same.

EXAM HINTS

There are several units to learn here. Take care not to confuse m/s (unit of speed and velocity) and m/s^2 (unit of acceleration).

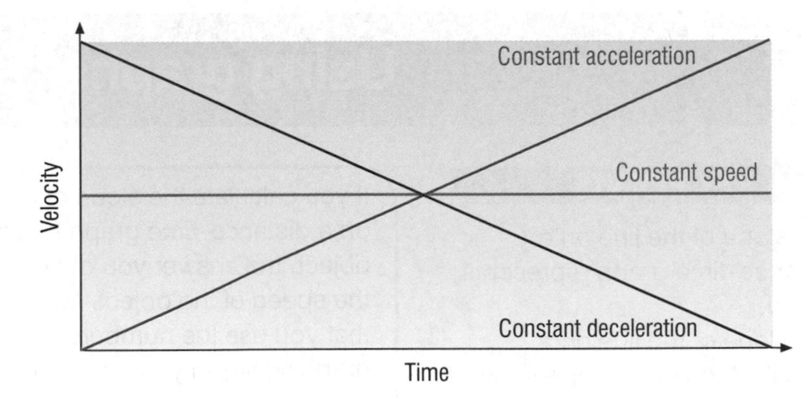

Velocity–time graphs

CHECK YOURSELF

1. What is the difference between speed and velocity?
2. What is the SI unit of acceleration?
3. What do we mean by deceleration?

students' book page 188

P2 1.3 More about velocity–time graphs

KEY POINTS

1. The slope of a line on a velocity–time graph represents acceleration.
2. The area under the line on a velocity–time graph represents distance travelled.

A velocity–time graph shows the velocity of a body (*y*-axis) against time taken (*x*-axis).

- The slope of a line on a velocity–time graph represents acceleration.
- The steeper the slope, the greater the acceleration.
- If the slope is negative, the body is decelerating.
- The area under the line on a velocity–time graph represents the distance travelled in a given time.
- The bigger the area, the greater the distance travelled.

Key words: area

AQA EXAMINER SAYS...

Take care not to confuse distance–time graphs and velocity–time graphs.

CHECK YOURSELF

1. What does a horizontal line on a velocity–time graph represent?
2. What would the velocity–time graph for a steadily decelerating body look like?
3. What does the area under the line on a velocity–time graph represent?

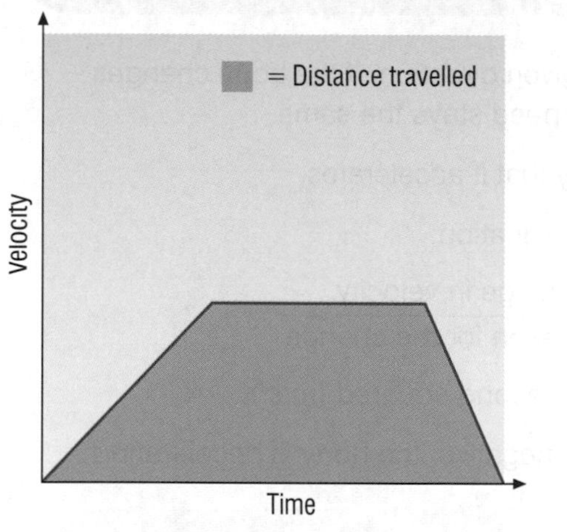

Area under a velocity–time graph

students' book page 190

P2 1.4 Using graphs

KEY POINTS

1. The slope of the line on a distance–time graph represents speed.
2. The slope of the line on a velocity–time graph represents acceleration.
3. The area under the line on a velocity–time graph represents the distance travelled. [Calculations of slopes are required for Higher Tier only]

HIGHER

If you calculate the slope of the line on a distance–time graph for an object, the answer you obtain will be the speed of the object. Make sure that you use the numbers from the graph scales in your calculations.

If you calculate the slope of the line on a velocity–time graph for an object, the answer you obtain will be the acceleration of the body. Make sure that you use the numbers from the graph scales in your calculations.

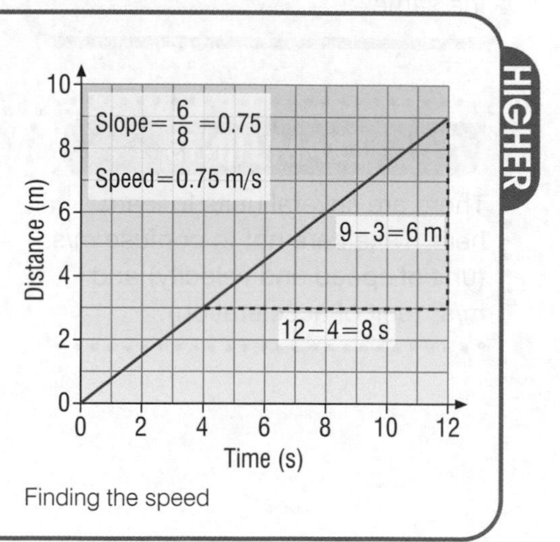

Finding the speed

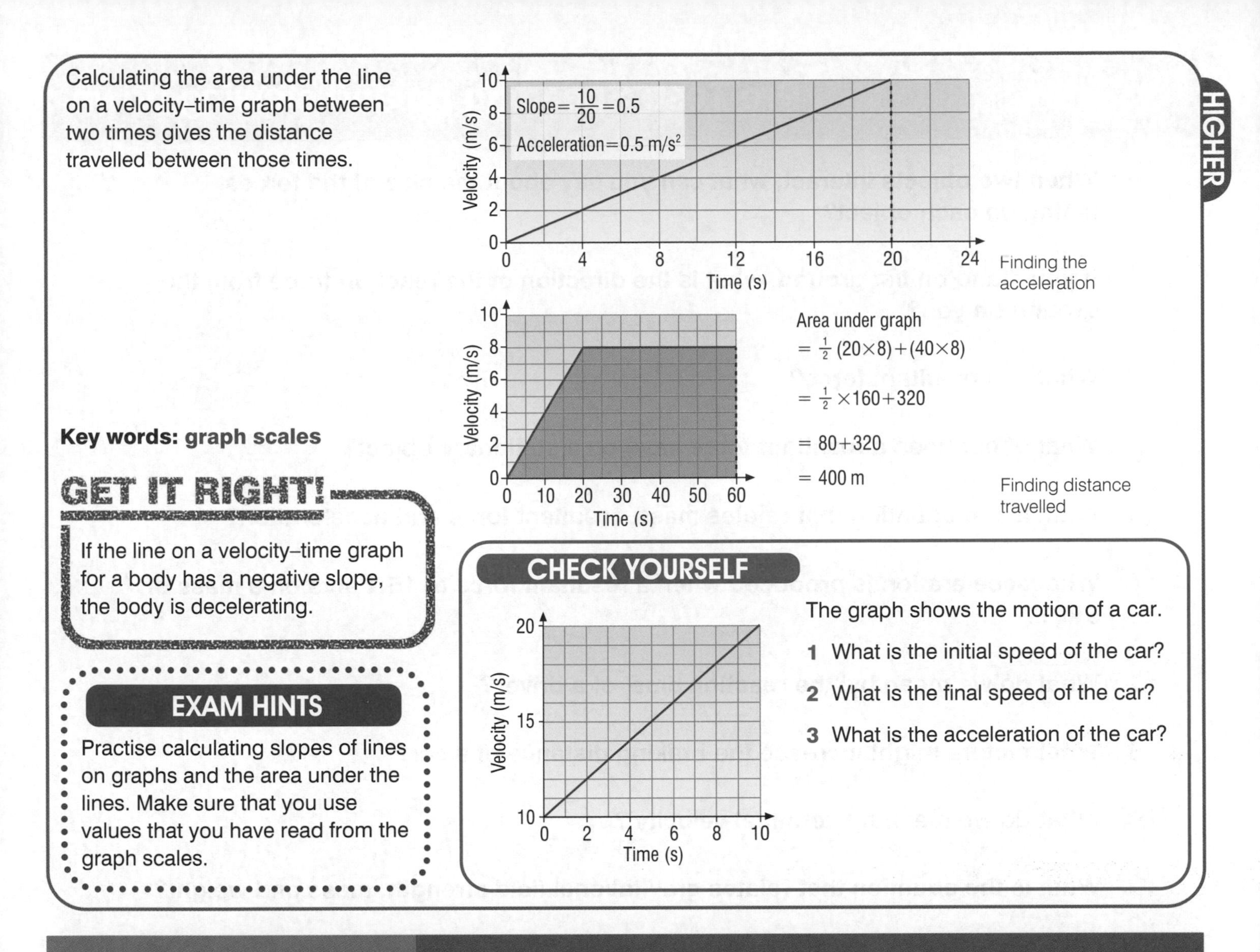

Calculating the area under the line on a velocity–time graph between two times gives the distance travelled between those times.

Key words: graph scales

GET IT RIGHT!

If the line on a velocity–time graph for a body has a negative slope, the body is decelerating.

EXAM HINTS

Practise calculating slopes of lines on graphs and the area under the lines. Make sure that you use values that you have read from the graph scales.

CHECK YOURSELF

The graph shows the motion of a car.

1 What is the initial speed of the car?

2 What is the final speed of the car?

3 What is the acceleration of the car?

P2 1 End of chapter questions

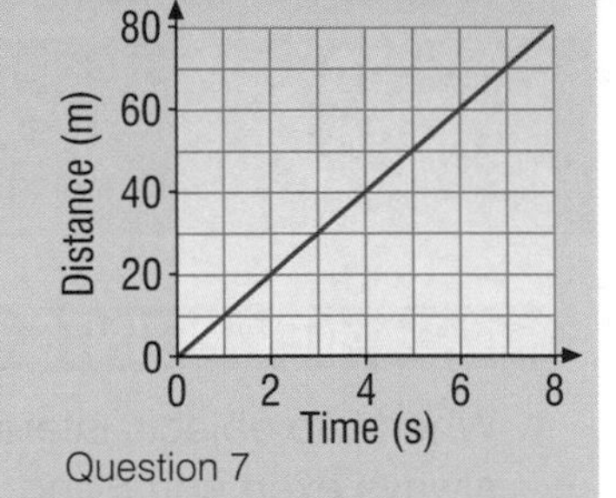

Question 7

1 **What does a horizontal line on a distance–time graph represent?**

2 **What is the average speed of a sprinter who runs 100 m in 16 s?**

3 **How can a body travelling at a steady speed be accelerating?**

4 **What quantity has the unit m/s^2?**

5 **What does a negative slope on a velocity–time graph mean?**

6 **What part of a velocity–time graph represents distance travelled?**

7 **Look at the top graph. What is the speed of the runner?**

8 **Look at the bottom graph. What is the distance travelled by the car in the first 10 seconds? [Higher Tier only]**

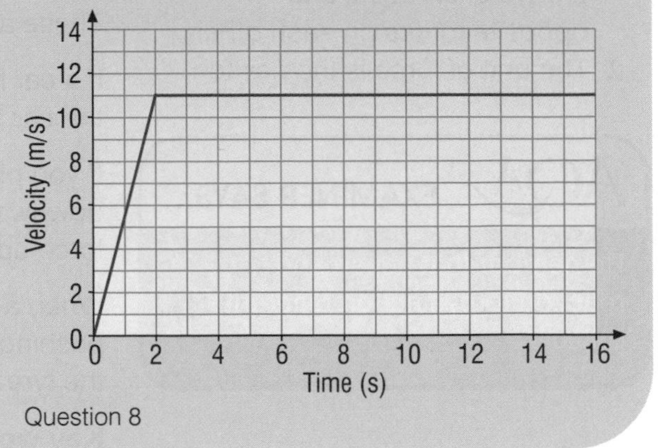

Question 8

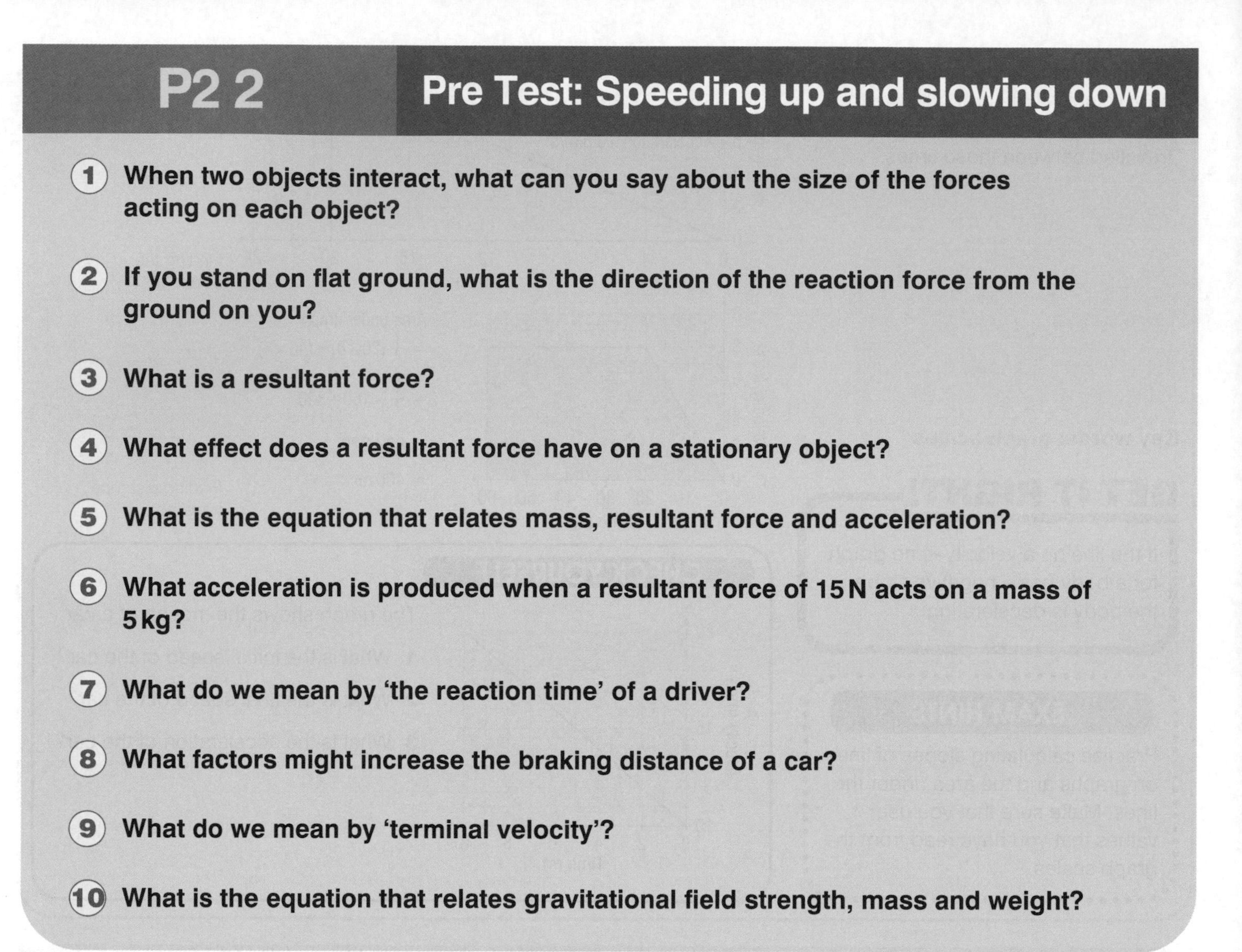

P2 2 Pre Test: Speeding up and slowing down

1. **When two objects interact, what can you say about the size of the forces acting on each object?**
2. **If you stand on flat ground, what is the direction of the reaction force from the ground on you?**
3. **What is a resultant force?**
4. **What effect does a resultant force have on a stationary object?**
5. **What is the equation that relates mass, resultant force and acceleration?**
6. **What acceleration is produced when a resultant force of 15 N acts on a mass of 5 kg?**
7. **What do we mean by 'the reaction time' of a driver?**
8. **What factors might increase the braking distance of a car?**
9. **What do we mean by 'terminal velocity'?**
10. **What is the equation that relates gravitational field strength, mass and weight?**

students' book page 196

P2 2.1 Forces between objects

KEY POINTS

1. When two objects interact, they always exert equal and opposite forces on each other.
2. The unit of force is the newton.

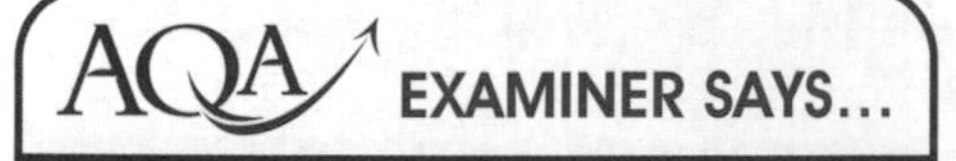

EXAMINER SAYS…

Remember that action and reaction forces act on different objects.

Forces are measured in newtons, abbreviated to N.

Objects always exert equal and opposite forces on each other. If object A exerts a force on object B, object B exerts an equal and opposite force on object A. These are sometimes called 'action and reaction' forces.

If a car hits a barrier it exerts a force on the barrier. The barrier exerts a force on the car that is equal in size and in the opposite direction.

If you place a book on a table the weight of the book will act vertically downwards on the table. The table will exert an equal and opposite reaction force upwards on the book.

When a car is being driven forwards there is a force from the tyre on the ground pushing backwards. There is an equal and opposite force from the ground on the tyre which pushes the car forwards.

Key words: force, newton, equal, opposite

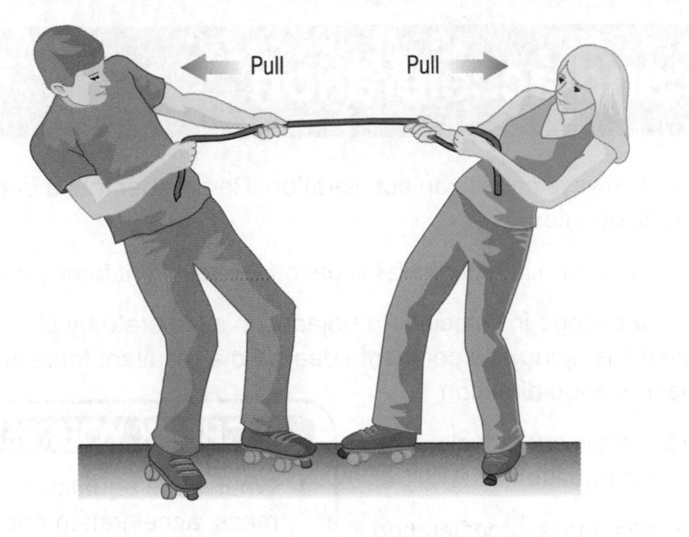

Equal and opposite forces

GET IT RIGHT!

Forces have both size and direction.

CHECK YOURSELF

1 What abbreviation is used for the unit of force?

2 In which direction does the force of weight always act?

3 If you push on a wall with a horizontal force of 15 N to the right, what force will the wall exert on you?

students' book page 198

P2 2.2 Resultant force

KEY POINTS

	Object at the start	Resultant force	Effect on the object
1	at rest	zero	stays at rest
2	moving	zero	velocity stays the same
3	moving	non-zero in the same direction as the direction of motion of the object	accelerates
4	moving	non-zero in the opposite direction to the direction of motion of the object	decelerates

Most objects have more than one force acting on them. The 'resultant force' is the single force that would have the same effect on the object as all the original forces acting together.

When the resultant force on an object is zero:

- if the object is at rest it will stay at rest
- if the object is moving it will carry on moving at the same speed and in the same direction.

When the resultant force on an object is not zero there will be an acceleration in the direction of the force.

This means that:

- if the object is at rest it will accelerate in the direction of the resultant force
- if the object is moving in the same direction as the resultant force it will accelerate in that direction
- if the object is moving in the opposite direction to the resultant force it will decelerate.

Key words: resultant force, accelerate, decelerate

GET IT RIGHT!

If an object is accelerating, there must be a resultant force acting on it.

If an object is accelerating, it can be speeding up, slowing down or changing direction.

CHECK YOURSELF

1 What happens to an object moving at a steady speed if the resultant force on it is zero?

2 What are the units of resultant force?

3 When will a resultant force cause a deceleration?

students' book page 200

P2 2.3 Force and acceleration

KEY POINTS

Resultant force (newtons, N) = mass (kilograms, kg) × acceleration (metres/second², m/s²)

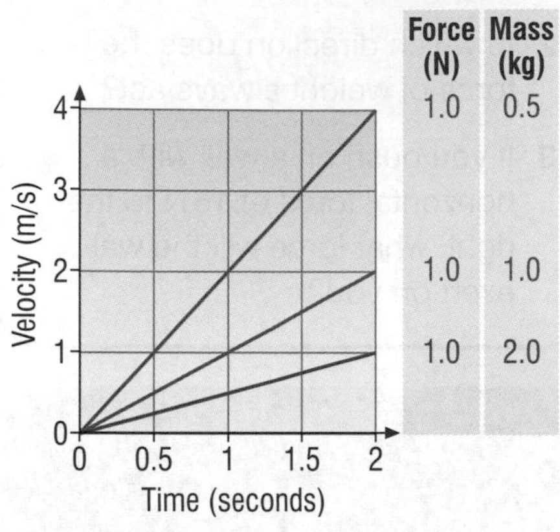

Different combinations of force and mass

A resultant force always causes an acceleration. Remember that a deceleration is a negative acceleration.

If there is no acceleration in a particular situation the resultant force must be zero.

Acceleration is a change in velocity. An object can accelerate by changing its direction even if it is going at a constant speed. So a resultant force is needed to make an object change direction.

Resultant force, mass and acceleration are related by the equation:

resultant force = mass × acceleration

The greater the resultant force on an object, the greater its acceleration.

The bigger the mass of an object, the bigger the force needed to give it a particular acceleration.

Key words: resultant force, mass, acceleration, deceleration

CHECK YOURSELF

1 What is the equation that links mass, acceleration and resultant force?

2 What happens to the acceleration of an object as the resultant force on it increases?

3 What is the acceleration of a car of mass 2000 kg if the resultant force acting in its direction of motion is 8 N?

students' book page 202

P2 2.4 On the road

KEY POINTS

1 The 'thinking' distance is the distance travelled by the vehicle in the time it takes the driver to react.
2 The braking distance is the distance the vehicle travels under the braking force.
3 Stopping distance = thinking distance + braking distance

GET IT RIGHT!

The reaction time depends on the driver. The braking distance depends on the road and weather conditions and the condition of the vehicle.

If a vehicle is travelling at a steady speed the resultant force on it is zero.

The driving forces are equal and opposite to the frictional forces.

The faster the speed of a vehicle, the bigger the deceleration needed to bring it to rest in a particular distance. So the bigger the braking force needed.

The stopping distance of a vehicle is the distance it travels during the driver's reaction time (the thinking distance) plus the distance it travels under the braking force (the braking distance).

The thinking distance is increased if the driver is tired or under the influence of alcohol or drugs.

The braking distance can be increased by poorly maintained roads, bad weather conditions and the condition of the car. For example, worn tyres or worn brakes will increase braking distance.

Key words: stopping distance, thinking distance, braking distance

CHECK YOURSELF

1 What is the resultant force on a car travelling at a steady speed on a straight horizontal road?

2 What is the relationship between stopping distance, thinking distance and braking distance?

3 What is the effect of the speed of a vehicle on its stopping distance?

students' book page 204

P2 2.5 Falling objects

KEY POINTS

1. The weight of an object is the force of gravity on it.
2. An object falling freely accelerates at about 10 m/s^2.
3. An object falling in a fluid reaches a terminal velocity.

GET IT RIGHT!

Do not confuse weight and mass. Remember that weight is the force of gravity acting on an object. Mass is the amount of matter in an object.

The resistive force exerted by a fluid is sometimes called the 'drag force'.

If an object falls freely, the resultant force acting on it is the force of gravity. It will make the object accelerate at about 10 m/s^2 close to the Earth's surface.

We call the force of gravity 'weight', and the acceleration 'the acceleration due to gravity'.

The equation resultant force = mass × acceleration

becomes weight (N) = mass (kg) × acceleration due to gravity (m/s^2)

If the object is on the Earth, not falling, we calculate the weight using:

weight (N) = mass (kg) × gravitational field strength (N/kg)

When an object falls through a fluid (e.g. air), the fluid exerts frictional forces (e.g. air resistance) on it, resisting its motion. The faster the object falls, the bigger the frictional force becomes. Eventually it will be equal to the weight of the object. The resultant force is now zero, so the body stops accelerating and moves at a constant velocity called the 'terminal velocity'.

We can show the motion of the object on a velocity–time graph.

Key words: weight, acceleration due to gravity, gravitational field strength, terminal velocity

CHECK YOURSELF

1. What are the units of gravitational field strength?
2. Why does an object dropped in a fluid initially accelerate?
3. What eventually happens to an object falling through a fluid?

P2 2 End of chapter questions

1. **If you pull on a rope attached to a wall with a force of 5 N, with what force will the rope pull on you?**
2. **When two objects interact, what can you say about the directions of the forces acting?**
3. **What effect does a resultant force have on the motion of an object moving in the same direction as the force?**
4. **What effect does a resultant force have on the motion of an object moving in the opposite direction to the force?**
5. **What happens to a body moving in a straight line if the resultant force acting on it increases?**
6. **What is the resultant force on a body of mass 70 kg accelerating at 2 m/s^2?**
7. **What do we mean by the 'thinking distance' of a car driver?**
8. **What factors might increase thinking distance?**
9. **What is the force of gravity acting on a object called?**
10. **What is the resultant force acting on a object at terminal velocity?**

P2 3 Pre Test: Work, energy and momentum

1. What is work done equivalent to?
2. What is the equation relating force, work done and distance?
3. What is the energy of movement called?
4. What is elastic potential energy?
5. What is the equation relating mass, velocity and momentum?
6. In what situations does the conservation of momentum apply?
7. A gun of mass 1 kg fires a bullet of mass 0.005 kg at 100 m/s. What is the recoil velocity of the gun?
8. A white snooker ball of mass 0.1 kg moving at 0.5 m/s hits an identical, stationary red ball. After the collision, the white ball continues in the same direction at a speed of 0.2 m/s. What is the speed of the red ball after the collision?
9. How does a crumple zone reduce the forces on a car in an accident?
10. A force of 2000 N is needed to bring a car to rest in 8 s. What is the momentum of the car?

students' book page 210

P2 3.1 Energy and work

KEY POINTS

1. Work done = energy transferred
2. Work done (joules, J) = force (newtons, N) × distance moved in the direction of the force (metres, m)

GET IT RIGHT!

Remember that work done is equal to energy transferred.

When a force moves an object, energy is transferred and work is done.

Whenever an object starts to move, a force must have been applied to it. This force needs a supply of energy from somewhere, such as electricity or fuel. When work is done moving the object, the supplied energy is transferred to the object so the work done is equal to the energy transferred.

Both work and energy have the unit joule, J.

When work is done against frictional forces, the energy supplied is mainly transformed into heat.

The work done on an object is calculated using the equation:

work done = force × distance moved in the direction of the force

Notice that if the distance moved is zero, no work is done on the object.

Key words: energy transferred, work done, joule

CHECK YOURSELF

1. What is the unit of work done?
2. When is work done by a force?
3. What work is done on an object if a force of 300 N moves it a distance of 7 m?

students' book page 212

P2 3.2 Kinetic energy

KEY POINTS

1 Elastic potential energy is the energy stored in an elastic object when work is done on the object.
2 The kinetic energy of an object depends on its mass and its speed.
3 Kinetic energy (joule, J) $= \frac{1}{2} \times$ mass (kilogram, kg) $\times$ speed^2 (metre/second)^2

BUMP UP YOUR GRADE

Take care when you use the kinetic energy equation. In an examination question students often forget to square the speed.

An elastic object is one that will go back to its original shape after it has been stretched or squashed.

When work is done on an elastic object to stretch or squash it, the energy transferred to it is stored as elastic potential energy. When the object returns to its original shape this energy is released.

Kinetic energy is the energy of movement.

The kinetic energy of a body depends on its mass and its speed. The greater its mass and the faster its speed, the more kinetic energy it has.

HIGHER

Kinetic energy can be calculated using the equation:

$$\text{kinetic energy} = \frac{1}{2}\text{mass} \times \text{speed}^2$$

CHECK YOURSELF

1 What happens to the kinetic energy of an object as its mass increases?
2 What are the units of kinetic energy?
3 What is the kinetic energy of a 1000 kg car travelling at 10 m/s? [Higher Tier only]

students' book page 214

P2 3.3 Momentum

KEY POINTS

1 Momentum (kg m/s) = mass (kg) × velocity (m/s)
2 Momentum is conserved whenever objects interact, as long as no external forces act on them.

GET IT RIGHT!

The units of momentum are kg m/s or N m.

A contact sport

All moving objects have momentum.

Momentum is calculated using the equation:

$$\text{momentum} = \text{mass} \times \text{velocity}$$

The units of momentum are kilogram metre/second, kg m/s.

Whenever objects interact, the total momentum before the interaction is equal to the total momentum afterwards, provided no external forces act on them.

This is called 'conservation of momentum'.

Another way to say this is that the total change in momentum is zero.

The interaction could be a collision or an explosion. After a collision the objects may move off together, or they may move apart.

Key words: mass, velocity, momentum

CHECK YOURSELF

1 When do objects have momentum?
2 What are the units of momentum?
3 What is the momentum of a 1000 kg car travelling at 10 m/s?

students' book page 216

P2 3.4 More on collisions and explosions

KEY POINTS

1 Momentum has size and direction.
2 When two objects push each other apart, they move apart with equal and opposite momentum.

Like velocity, momentum has both size and direction. In calculations one direction must be defined as positive, so momentum in the opposite direction is negative.

When two objects are at rest their momentum is zero. In an explosion the objects move apart with equal and opposite momentum. One momentum is positive and the other negative, so the total momentum after the explosion is zero.

Firing a bullet from a gun is an example of an explosion. The bullet moves off with a momentum in one direction and the gun 'recoils' with equal momentum in the opposite direction.

Key words: direction, positive, negative, equal, opposite, explosion

GET IT RIGHT!

The total momentum before a collision is usually zero. So the total momentum afterwards will be zero.

EXAM HINTS

In calculations it often helps to sketch a quick diagram to show what the objects are doing before and after the collision.

CHECK YOURSELF

1 What must the total momentum after an explosion be equal to?
2 Two physics students on roller skates stand holding each other in the playground. They push each other away. What can you say about the momentum of each student?
3 One student has twice the mass of the other. What can you say about the velocity of each student?

students' book page 218

P2 3.5 Changing momentum

KEY POINTS

1 The more time an impact takes, the less the force exerted.
2 Force (N) = $\frac{\text{change of momentum (kg m/s)}}{\text{time taken for change (s)}}$
[Higher Tier only]

HIGHER

When a force acts on an object that is moving, or able to move, its momentum changes. The equation that describes this is:

$$\text{force} = \frac{\text{change in momentum}}{\text{time taken for the change}}$$

For a particular change in momentum, the longer the time taken for the change, the smaller the force.

In a collision, the momentum of an object often becomes zero during the impact – the object comes to rest. If the impact time is short, the forces on the object are large. As the impact time increases, the forces become less.

This idea is the basis of a number of safety features in cars. Crumple zones in cars are designed to fold in a collision. This increases the impact time and so reduces the force on the car and the people in it.

Air bags work in a similar way. The driver's head changes momentum slowly when it hits an airbag. So the force on the head is less than if it changes momentum quickly by hitting the steering wheel.

Key words: impact time, change in momentum, safety features

AQA EXAMINER SAYS...

Force multiplied by time taken to change momentum is called the impulse.

BUMP UP YOUR GRADE

Make sure that you can explain how crumple zones and air bags reduce the forces acting by increasing the time taken to change the momentum of a car.

A crash test

CHECK YOURSELF

1 What is the equation that relates force, change in momentum and time? [Higher Tier only]

2 A car has a momentum of 30 000 kg m/s. Calculate the force needed to stop the car in 12 s. [Higher Tier only]

3 Why is a gymnast less likely to injure herself if she lands on a thick foam mat than if she lands on a hard floor?

P2 3 End of chapter questions

1 What are the units of energy transferred?

2 What force must be applied to an object if 2800 J of work are done moving it 7 m?

3 What is the equation that relates mass, kinetic energy and speed? [Higher Tier only]

4 What is the mass of a car that has 4000 J of kinetic energy when moving at 10 m/s? [Higher Tier only]

5 What is the momentum of a 2000 kg truck when it is travelling at 20 m/s?

6 A trolley of mass 0.2 kg moving at 1.5 m/s to the right collides with a stationary trolley of mass 0.3 kg. After the collision they move off together. Calculate the velocity of the trolleys after the collision.

7 What is the total momentum after a collision equal to?

8 Why might you calculate a value for velocity that is negative?

9 A hammer of mass 1 kg hits a nail when travelling at 0.5 m/s. It takes 0.2 s to come to rest. What is the force applied to the nail? [Higher Tier only]

10 A force of 2000 N is needed to bring a car to rest in 8 s. What is the initial momentum of the car? [Higher Tier only]

EXAMINATION-STYLE QUESTIONS

1 The diagram shows the forces acting on a flying helicopter.

(a) What is the name of:
  (i) force X
  (ii) force Y?
  (2 marks)

(b) Describe the motion of the helicopter, if the force X is equal in size to the lift force and force Y is equal in size to the thrust force. (2 marks)

(c) Which force must decrease to make the helicopter:
  (i) decrease its height
  (ii) decrease its forward speed? (2 marks)

2 The diagram shows a man pushing a box up a slope with a force of 200 N.

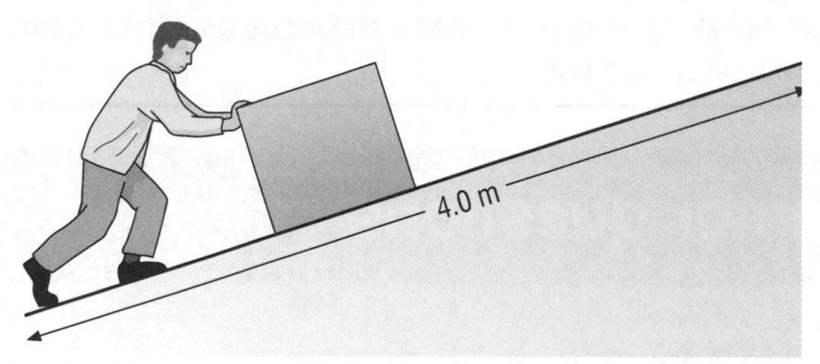

(a) The mass of the box is 50 kg. The gravitational field strength is 10 N/kg. Calculate the weight of the box. (2 marks)

(b) Calculate the work done pushing the box up the slope. (2 marks)

3 The drawing shows a ball being hit by a cricket bat.

(a) The ball has a mass of 0.16 kg. The bat is in contact with the ball for 0.02 s. The force exerted by the bat on the ball is 200 N.
  (i) Write down the equation that links change in momentum, force and time. (1 mark)
  (ii) Calculate the velocity of the ball, in m/s, away from the bat. (4 marks)

(b) A bowler bowls the ball with a speed of 25 m/s.

Calculate, in joules, the kinetic energy of the ball. (3 marks)

[Higher]

4 A student is investigating the conservation of momentum. She is using a horizontal air track and two gliders.

(a) What is the *conservation of momentum*? (2 marks)

(b) In what types of event does the conservation of momentum apply? (2 marks)

(c) The diagram shows the air track and the two gliders, X and Y.

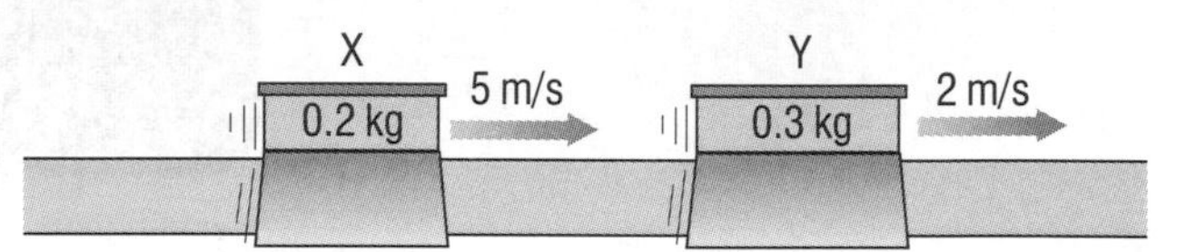

The mass of X is 0.2 kg and its velocity is 5 m/s. It collides with Y, which has a mass of 0.3 kg and is moving in the same direction at 2 m/s. The two gliders join together.

Calculate the speed of the gliders after the collision. (4 marks)

5 The graph below shows how the thinking distance and braking distance for a car increase with speed.

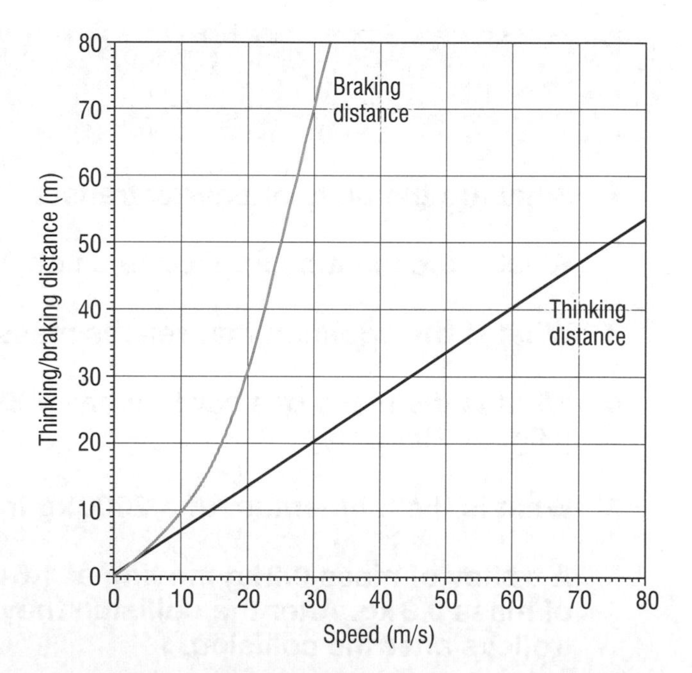

(a) Comment on how thinking distance and braking distance increase with speed. (2 marks)

(b) What is the total stopping distance at 25 m/s? (4 marks)

(c) What is the effect on thinking distance and braking distance of:
  (i) driving on a wet road surface? (2 marks)
  (ii) driving when very tired? (2 marks)

The answer is worth 4 of the 6 marks available.

The responses worth a mark are underlined in red.

We can improve the answer in several ways:

Many cars are fitted with seat belts and driver airbags. Use your understanding of momentum to explain why airbags and seat belts reduce injuries to the driver in the event of a high speed crash. *(6 marks)*

Change in momentum = force × time

The airbag increases the time taken for the momentum change.

So force on the driver is less

So there is less chance of serious injury

The question asks about airbags and seatbelts, but seatbelts are not mentioned.

To gain the other 2 marks state that a driver going at a high speed will have a large momentum, and so a large change in momentum when they stop.

The answer is worth 5 of the 8 marks available for this Higher Tier question.

The responses worth a mark are underlined in red.

We can improve the answer in several ways:

A car travels on a straight, level road. The graph shows how the speed of the car changes with time.

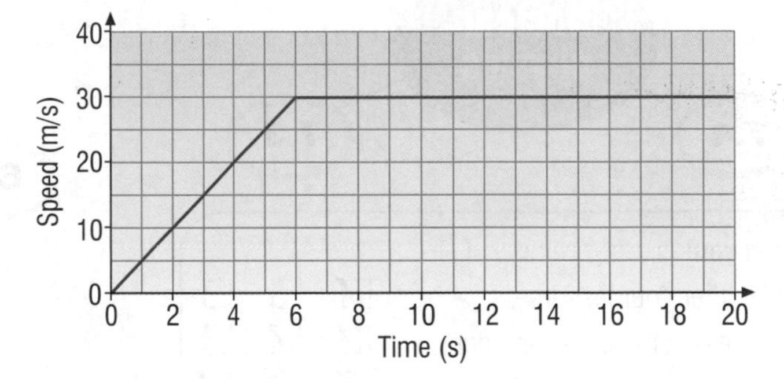

(a) Calculate the acceleration of the car during the first 6 seconds. *(4 marks)*

(b) Calculate the distance travelled by the car during the first 20 seconds of the journey. *(4 marks)*

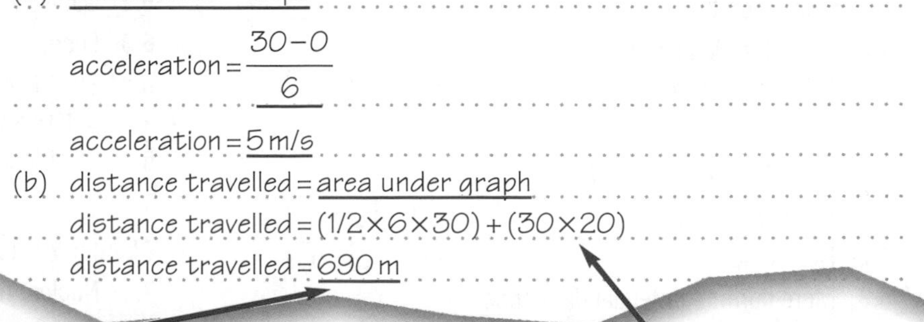

(a) acceleration = slope

acceleration = $\frac{30-0}{6}$

acceleration = 5 m/s

(b) distance travelled = area under graph

distance travelled = (1/2 × 6 × 30) + (30 × 20)

distance travelled = 690 m

An incorrect unit is given in the answer. The units for speed (m/s) and acceleration (m/s^2) have been mixed up, so no final mark.

The final answer is wrong but the unit is correct, so one of the final 2 marks is given.

The length of the rectangle is incorrectly written as 20, not 14.

P2 | Additional physics (Chapters 4–7)

Checklist

This spider diagram shows the topics in the second half of the unit. You can copy it out and add your notes and questions around it, or cross off each section when you feel confident you know it for your exams.

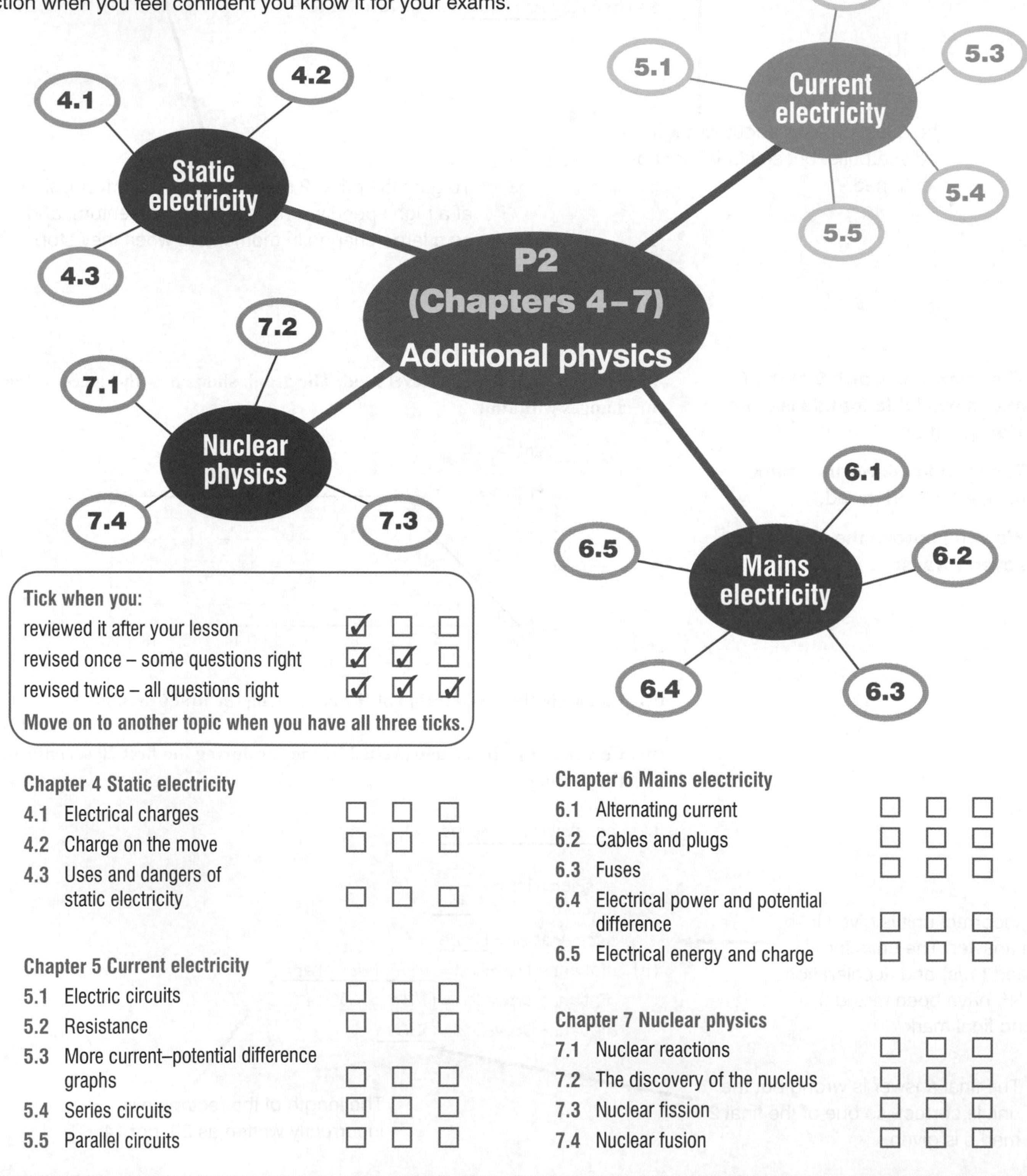

Chapter 4 Static electricity
- 4.1 Electrical charges ☐ ☐ ☐
- 4.2 Charge on the move ☐ ☐ ☐
- 4.3 Uses and dangers of static electricity ☐ ☐ ☐

Chapter 5 Current electricity
- 5.1 Electric circuits ☐ ☐ ☐
- 5.2 Resistance ☐ ☐ ☐
- 5.3 More current–potential difference graphs ☐ ☐ ☐
- 5.4 Series circuits ☐ ☐ ☐
- 5.5 Parallel circuits ☐ ☐ ☐

Chapter 6 Mains electricity
- 6.1 Alternating current ☐ ☐ ☐
- 6.2 Cables and plugs ☐ ☐ ☐
- 6.3 Fuses ☐ ☐ ☐
- 6.4 Electrical power and potential difference ☐ ☐ ☐
- 6.5 Electrical energy and charge ☐ ☐ ☐

Chapter 7 Nuclear physics
- 7.1 Nuclear reactions ☐ ☐ ☐
- 7.2 The discovery of the nucleus ☐ ☐ ☐
- 7.3 Nuclear fission ☐ ☐ ☐
- 7.4 Nuclear fusion ☐ ☐ ☐

What are you expected to know?

Chapter 4 Static electricity (See students' book pages 224–233)

- When insulating materials are rubbed against each other they may become charged.
- Opposite charges attract, like charges repel.
- Electrostatic charge can be both useful and dangerous.

Chapter 5 Current electricity (See students' book pages 234–247)

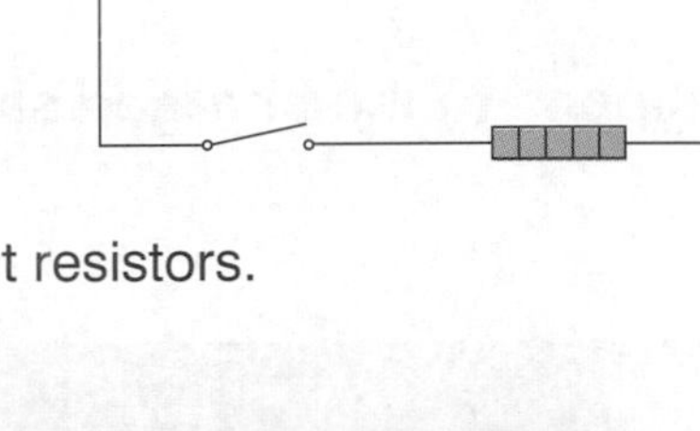

- Common circuit symbols.
- Current–potential difference graphs for resistor, lamp and diode.
- Ohm's law: Potential difference = current × resistance
- Using Ohm's law with series and parallel circuits.
- Variation in the resistance of thermistors and light dependent resistors.

Chapter 6 Mains electricity

(See students' book pages 248–261)

- Correct wiring of a three-pin plug.
- How fuses are used.
- Charge = current × time
- Power = current × potential difference
- Energy transferred = potential difference × charge

Chapter 7 Nuclear physics (See students' book pages 262–273)

- How the Rutherford–Marsden scattering experiment lead to the nuclear model of the atom.
- Relative charges and masses of protons, neutrons and electrons.
- The meaning of atomic number, mass number and isotope.
- Origins of background radiation.
- The effect of alpha and beta decay on nuclei.
- The process of fission and the meaning of a chain reaction.
- The process of fusion.

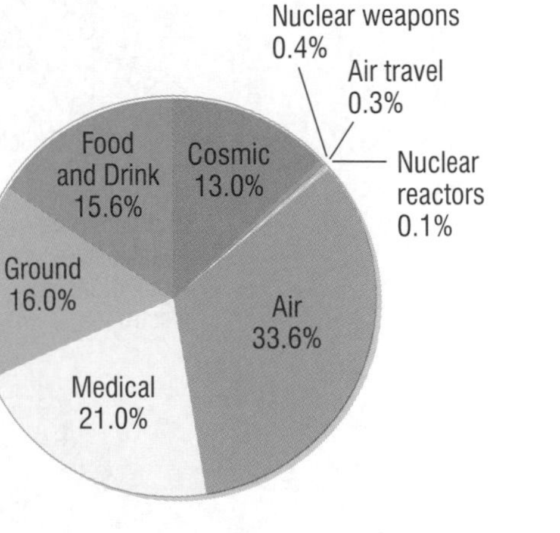

P2 4 Pre Test: Static electricity

1. How does an insulator become negatively charged?
2. What will happen if two negatively charged objects are brought close to each other?
3. What is meant by a 'conductor' of electricity?
4. Why are metals good conductors?
5. In an electrostatic smoke precipitator, how do the smoke particles become charged?
6. What happens to the charged smoke particles?

students' book page 224

P2 4.1 Electrical charges

KEY POINTS

1. Like charges repel; unlike charges attract.
2. Insulating materials that lose electrons when rubbed become positively charged.
3. Insulating materials that gain electrons when rubbed become negatively charged.

When two electrically insulating materials are rubbed together electrons are rubbed off one material and deposited on the other. Which way the electrons are transferred depends on the particular materials used.

Electrons have a negative charge so the material that has gained electrons becomes negatively charged. The one that has lost electrons is left with a positive charge.

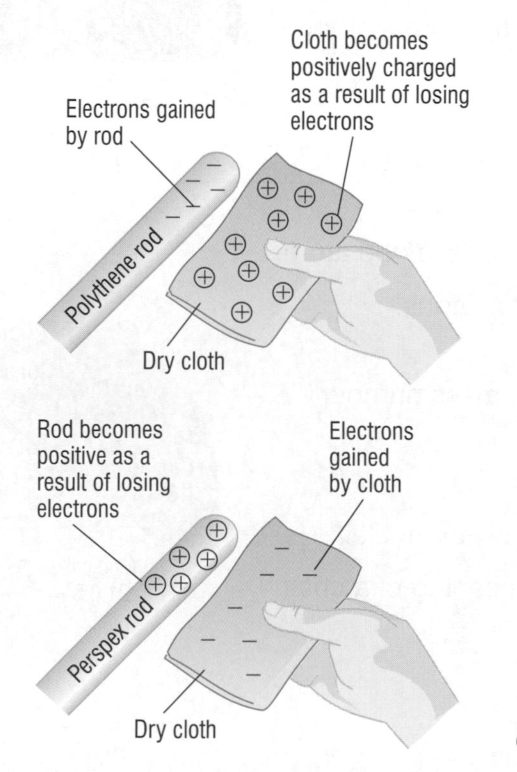

Charging by friction

GET IT RIGHT!

It is only ever electrons that move to produce positive and negative static charges on objects.

Two objects that have opposite electric charges are attracted to each other. Two objects that have the same electric charges repel each other. The bigger the distance between the objects, the weaker the force.

Key words: insulating, electrons, negative, positive, static, attract, repel

CHECK YOURSELF

1 What sort of charge does an electron have?

2 How does an insulator become positively charged?

3 What sort of force will there be between two negatively charged objects?

students' book page 226

P2 4.2 Charge on the move

KEY POINTS

1 Electric current is the rate of flow of charge.
2 A metal object can only hold charge if it is isolated from the ground.
3 A metal object is earthed by connecting it to the ground.

When charge flows through a conductor there is a current in it. Electric current is the rate of flow of charge.

In a solid conductor, e.g. a metal wire, the charge carriers are electrons.

Metals are good conductors of electricity because they contain free, conduction electrons that are not confined to a single atom.

Insulators cannot conduct because all the electrons are held in atoms.

A conductor can only hold charge if it is isolated from the ground. Otherwise electrons will flow to or from the earth and discharge it.

HIGHER

The bigger the charge on an isolated object, the higher the potential difference between the object and the earth. If the potential difference becomes high enough, a spark may jump across the gap between the object and any earthed conductor brought near it.

GET IT RIGHT!

If a negatively charged, isolated conductor is earthed, electrons will flow from it to the earth until it is no longer charged. It the conductor is positively charged, electrons will flow from the earth to it.

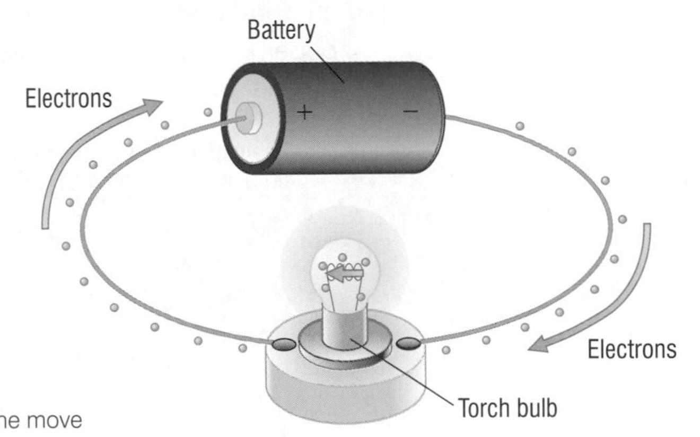

Electrons on the move

CHECK YOURSELF

1 Why can't insulators conduct?

2 How can a metal conductor be made to hold charge?

3 What is electric current?

students' book page 228

P2 4.3 Uses and dangers of static electricity

KEY POINTS

1 A spark from a charged object can make powder grains or certain gases explode.
2 To eliminate static electricity:
- use antistatic materials
- earth metal pipes and objects.

EXAM HINTS

Make sure you understand these uses and dangers of static electricity.

In the exam you might be asked to explain one of them in detail.

In some situations electrostatic charge can be useful, and in some it can be dangerous.

Objects such as cars panels and bicycle frames are often painted with an electrostatic paint sprayer. The spray nozzle is connected to a positive terminal. As the paint droplets pass through it, they pick up a positive charge. The paint drops repel each other so they spread out to form a fine cloud. The item being painted is connected to a negative terminal so the positively charged droplets are attracted to it.

An electrostatic paint sprayer

In a photocopier, a copying plate is given a charge. An image of the page to be copied is projected onto the charged plate. Where light hits the plate the charge leaks away, leaving a pattern of the page. Black ink powder is attracted to the charged parts of the plate. This powder is transferred onto a piece of paper. The paper is heated so the powder melts and sticks to it, producing a copy of the original page.

1 Photocopiers with a photoconducting drum – drum positively charged until light falls on it.

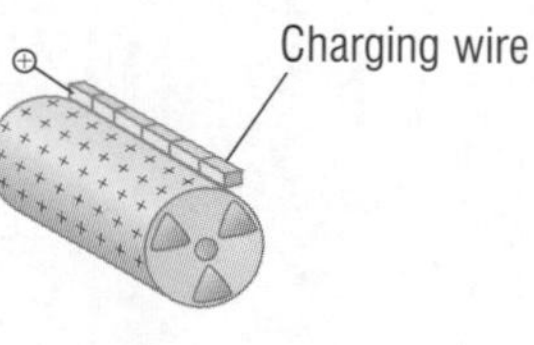

2 Light reflected off the paper onto the drum. The areas of black do not reflect so the drum keeps its charge in these areas.

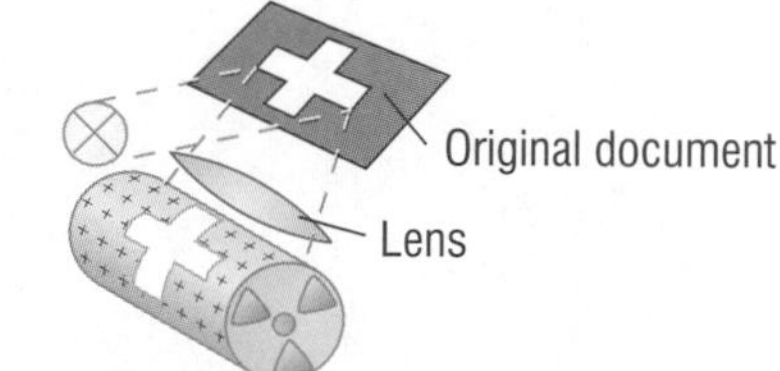

3 The black toner sticks to the drum where it is still charged and is pressed onto paper.

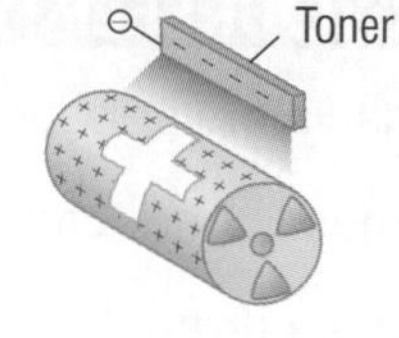

4 The paper is finally heated to stick the toner to it permanently.

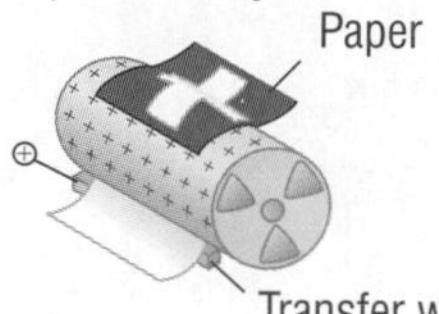

Inside a photocopier

EXAM HINTS

Don't panic if you see a question in the exam about a use or danger of static electricity that you haven't seen before.

In these questions, you are expected to apply what you know about static electricity to an unfamiliar situation.

Electrostatic smoke precipitators are used in chimneys to attract dust and smoke particles so that they are not released into the air. The particles pass a charged grid and pick up charge. They are then attracted to plates on the chimney walls that have the opposite charge. The particles stick to the plates until they are shaken off and collected.

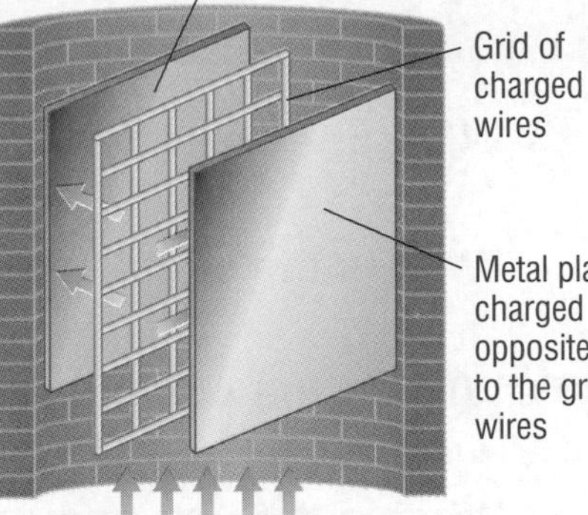

An electrostatic precipitator

When powder or grain flows through a pipe, friction makes it pick up charge. This could lead to a spark igniting the powder, causing an explosion.

The filler pipes on road tankers that are used to pump fuel into storage tanks are earthed to prevent them becoming charged. A spark could cause an explosion of the fuel vapour.

Key words: electrostatic paint sprayer, photocopier, smoke precipitator

CHECK YOURSELF

1 Why are the filler pipes on fuel tankers earthed?

2 Why is the spray nozzle of an electrostatic paint sprayer connected to a positive terminal?

3 How does grain pick up charge as it flows through a pipe?

P2 4 End of chapter questions

1 Why do objects become positively or negatively charged only by the movement of negative charges?

2 Why do you sometimes get a shock from synthetic clothing when you take it off?

3 What is a lightning conductor made of?

4 How can an isolated, charged conductor be discharged?

5 Give three examples of devices that make use of static electricity.

6 A moving car may become charged. What would happen if you touched the door of the car?

P2 5 Pre Test: Current electricity

1. What does a battery consist of?
2. What is the circuit symbol for a fuse?
3. What is the equation relating current, potential difference and resistance?
4. What is the unit of current?
5. What is the circuit symbol for a thermistor?
6. What happens to the resistance of a thermistor if its temperature increases?
7. How would you calculate the current in a series circuit?
8. Why is the current through each component in a series circuit the same?
9. Why is the potential difference (p.d.) across each component connected in parallel the same?
10. How would you calculate the total current in a parallel circuit?

students' book page 234

P2 5.1 Electric circuits

KEY POINTS

1. Every component has its own agreed symbol.
2. A circuit diagram shows how components are connected together.
3. A battery consists of two or more cells connected together.

Every electrical component has an agreed symbol. Some of them are shown below:

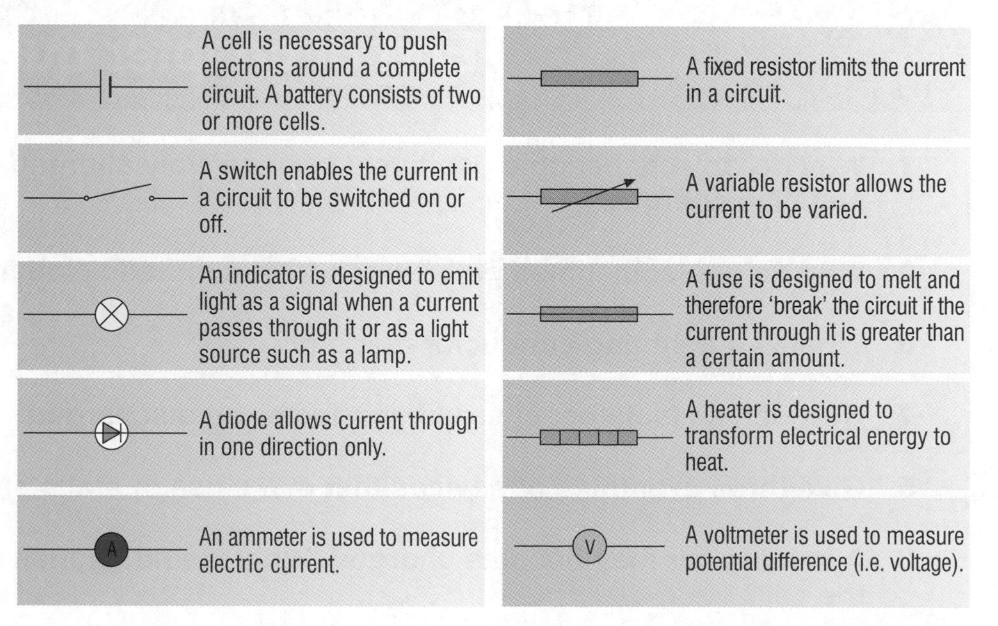

Components and symbols

EXAMINER SAYS...

Make sure you can recognise and draw all of these circuit symbols.

Use a sharp pencil and make sure there are no gaps or odd ends of wire shown on your diagram.

GET IT RIGHT!

A battery consists of two or more cells connected together.

A circuit diagram uses the symbols to show how components are connected together to make a circuit.

Key words: circuit symbol

CHECK YOURSELF

1 What is the circuit symbol for a variable resistor?

2 What is the circuit symbol for a diode?

3 Draw a circuit diagram for a circuit containing a cell, a lamp, a resistor and a switch connected one after the other.

students' book page 236

P2 5.2 Resistance

KEY POINTS

1 Resistance (ohms, Ω) = $\frac{\text{potential difference (volts, V)}}{\text{current (amperes, A)}}$

2 The current through a resistor at constant temperature is directly proportional to the potential difference across the resistor.

GET IT RIGHT!

A graph with a straight line that passes through the origin shows that the two variables are directly proportional to each other.

EXAMINER SAYS...

Remember that the resistor must stay at a constant temperature or the graph will not be a straight line.

Current–potential difference graphs are used to show how the current through a component varies with the potential difference across it.

The current is measured with an ammeter. Ammeters are always placed in series with the component. The unit of current is the ampere, A.

The potential difference is measured with a voltmeter. Voltmeters are always placed in parallel with the component. The unit of potential difference is the volt, V.

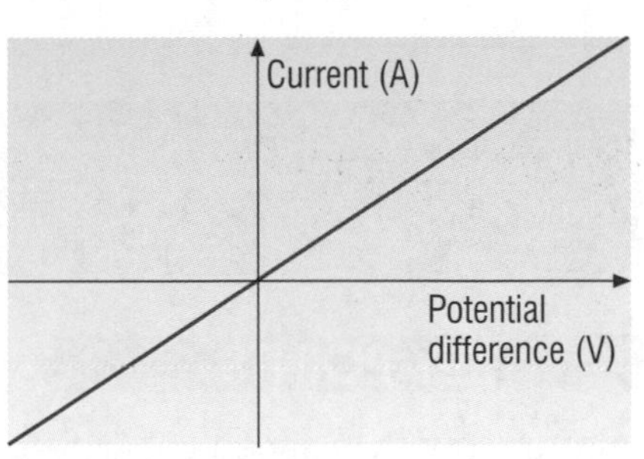

A current–p.d. graph for a wire at constant temperature

If the resistor is kept at a constant temperature the graph shows a straight line passing through the origin. This means the current is directly proportional to the potential difference across the resistor.

Potential difference and current are related by an equation called Ohms law:

potential difference = current × resistance.

Resistance is measured in ohms, Ω, and is the opposition to charge flowing through the resistor. If the resistor is kept at a constant temperature, the resistance stays constant. If the temperature of the resistor increases, the resistance increases. So the line on the current–p.d. graph is no longer straight.

Key words: current, potential difference, resistance

CHECK YOURSELF

1 What device is used to measure current?

2 What is the unit of resistance?

3 What do we mean by 'resistance'?

students' book page 238

P2 5.3 More current–potential difference graphs

KEY POINTS

1. In a filament lamp, resistance increases with increase of the filament temperature.
2. In a diode, 'forward' resistance is low, 'reverse' resistance is high.
3. In a thermistor, resistance decreases if its temperature increases.
4. In an LDR, resistance decreases if the light intensity on it increases.

The current–potential difference graph for a filament lamp curves. So the current is not directly proportional to the potential difference.

The resistance of the filament increases as the current increases. This is because the resistance increases as the temperature increases.

Reversing the potential difference makes no difference to the shape of the curve.

The current through a diode flows in one direction only. In the reverse direction the diode has a very high resistance so the current is zero.

As the light falling on it gets brighter, the resistance of a light-dependent resistor (LDR) decreases.

As the temperature goes up, the resistance of a thermistor goes down.

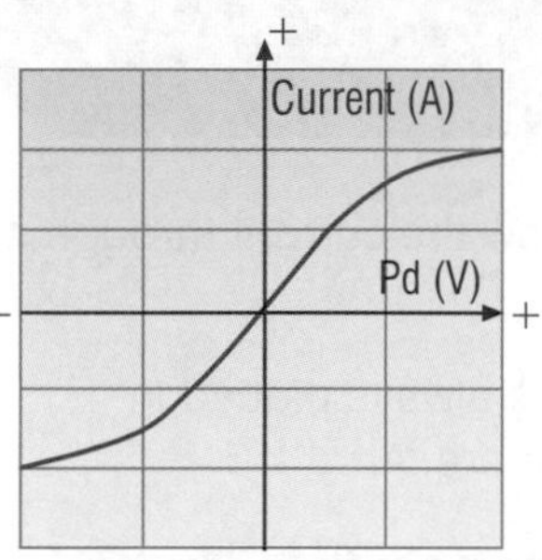

A current–potential difference graph for a filament lamp

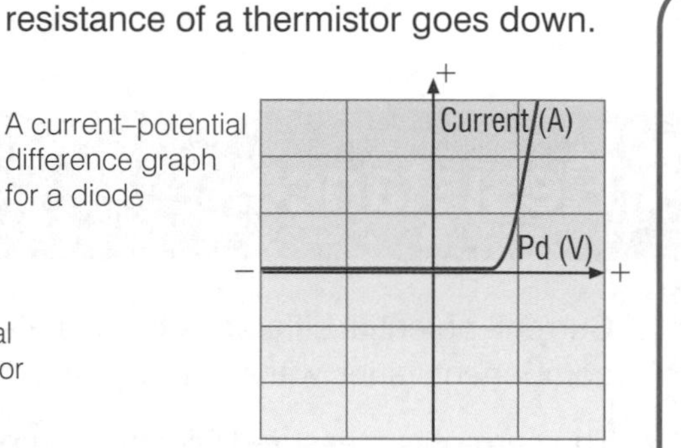

A current–potential difference graph for a diode

Key words: filament lamp, diode, light dependent resistor, thermistor

CHECK YOURSELF

1. What happens to the resistance of a LDR if its surroundings become darker?
2. What effect does reversing the potential difference across a filament lamp have?
3. Explain the shape of a current–potential difference graph for a diode.

students' book page 240

P2 5.4 Series circuits

KEY POINTS

For components in series:

1. The current is the same in each component.
2. The potential differences add to give the total potential difference.
3. The resistances add to give the total resistance.

In a series circuit the components are connected one after another, so if there is a break anywhere in the circuit charge stops flowing. There is no choice of route for the charge as it flows around the circuit so the current through each component is the same.

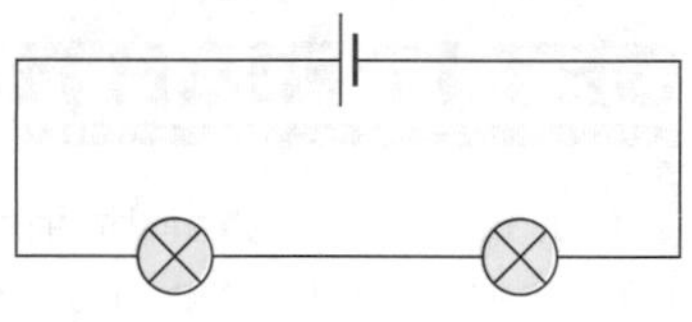
Lamps in series

The current depends on the potential difference (p.d.) of the supply and the total resistance of the circuit:

$$\text{current} = \frac{\text{p.d. of supply}}{\text{total resistance}}$$

The p.d. of the supply is shared between all the components in the circuit. So the p.d.s across individual components add up to give the p.d. of the supply.

The resistances of the individual components in series add up to give the total resistance of the circuit.

The bigger the resistance of a component, the bigger its share of the supply p.d.

Key words: series, supply

GET IT RIGHT!

Remember that in a series circuit the current is the same everywhere in the circuit:

$$\text{potential difference} = \text{current} \times \text{resistance}$$

This can be used to find the potential difference across each individual component if its resistance is known.

CHECK YOURSELF

1. What happens in a series circuit if one component stops working?
2. How could you find the total resistance in a series circuit?
3. A series circuit contains a variable resistor. If its resistance is increased, what happens to the p.d. across it?

students' book page 242

P2 5.5 Parallel circuits

KEY POINTS

For components in parallel:

1. The potential difference is the same across each component.
2. The total current is the sum of the currents through each component.
3. The bigger the resistance of a component, the smaller its current is.

In a parallel circuit each component is connected across the supply, so if there is a break in one part of the circuit charge can still flow in the other parts.

Each component is connected across the supply p.d., so the p.d. across each component is the same.

Lamps in parallel

There are junctions in the circuit so different amounts of charge can flow through different components. The current through each component depends on its resistance. The bigger the resistance of a component, the smaller the current through it.

The total current through the whole circuit is equal to the sum of the currents through the separate components.

Key words: parallel

GET IT RIGHT!

In everyday life parallel circuits are much more useful than series circuits, as a break in one part of the circuit does not stop current flowing in the rest of the circuit.

EXAMINER SAYS...

Make sure that you understand the differences between series and parallel circuits.

CHECK YOURSELF

1. What happens in a parallel circuit if one component stops working?
2. In a parallel circuit what is the relationship between the supply p.d. and the p.d. across each parallel component?
3. How can you find the total current in a parallel circuit?

P2 5 End of chapter questions

1. **What is a circuit diagram?**
2. **Draw a circuit diagram for a circuit containing a battery, a resistor, a variable resistor, a switch and an ammeter.**
3. **Where is an ammeter placed in a circuit?**
4. **What device is used to measure potential difference?**
5. **What is the circuit symbol for a light dependent resistor?**
6. **Why does the current–potential difference graph for a filament lamp curve?**
7. **A 2 Ω, a 6 Ω and a 10 Ω resistor are placed in series. What is their total resistance?**
8. **The total resistance in a series circuit is 24 Ω and the p.d. of the supply is 12 V. What is the current in the circuit?**
9. **A parallel circuit contains a variable resistor. If its resistance increases, what happens to the p.d. across it?**
10. **A parallel circuit contains a variable resistor. If its resistance increases, what happens to the current through it?**

P2 6 Pre Test: Mains electricity

1. What is alternating current?
2. What is the frequency of the UK mains supply?
3. What colour is the live wire?
4. What is the outer cover of a three-pin plug made of?
5. What does a fuse contain?
6. What will happen if the live wire touches the metal case of an appliance?
7. What is the equation that relates power, energy and time?
8. What is the unit of electrical power?
9. What is the equation that relates charge, current and time?
10. What energy transformation takes place when charge flows through a resistor?

students' book page 248

P2 6.1 Alternating current

KEY POINTS

1. Alternating current repeatedly reverses its direction.
2. Mains electricity is an alternating current supply.
3. A mains circuit has a live wire which is alternately positive and negative every cycle and a neutral wire at zero volts. [Higher Tier only]

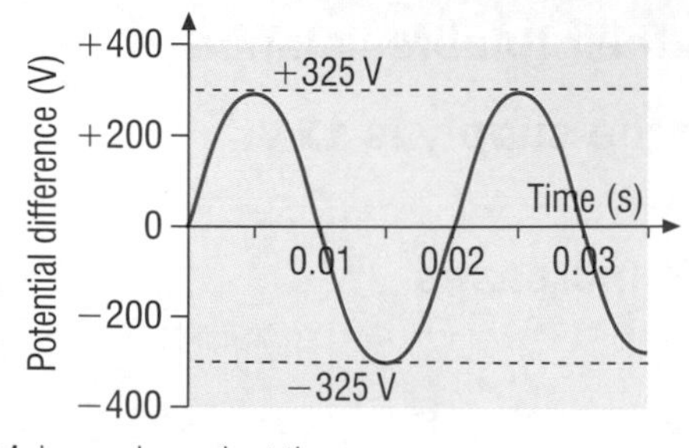

Mains p.d. against time

Cells and batteries supply current that passes round the circuit in one direction. This is called direct current, or d.c.

The current from the mains supply passes in one direction, then reverses and passes in the other direction. This is called alternating current, or a.c.

The frequency of the UK mains supply is 50 Hertz, which means it alternates direction 50 times each second. The 'voltage' of the mains is 230 V.

HIGHER

The live wire of the mains supply alternates between a positive and a negative potential with respect to the neutral wire. The neutral wire stays at zero volts.

The live wire alternates between +325 volts and −325 volts. In terms of electrical power, this is equivalent to a direct potential difference of 230 volts.

Key words: direct current, alternating current, live, neutral

EXAM HINTS

Make sure you can take readings from diagrams of an oscilloscope trace.

CHECK YOURSELF

1. What is direct current?
2. What is the potential of the neutral terminal? [Higher Tier only]
3. What is the potential difference of the mains supply?

students' book page 250

P2 6.2 Cables and plugs

KEY POINTS

1 Cables consist of two or three insulated copper wires surrounded by an outer layer of flexible plastic material.
2 Sockets and plugs are made of stiff plastic materials, which enclose the electrical connections.
3 In a three-pin plug or a three-core cable:
   - The live wire is brown.
   - The neutral wire is blue.
   - The earth wire is yellow/green.
   - The earth wire is used to earth the metal case of a mains appliance.

Mains cable

EXAM HINTS

Make sure you can identify faults in the wiring of a three-pin plug.

Mains electricity can be dangerous unless used with care. Avoid hazards such as:

- Overlong or frayed cables.
- Using electricity near sources of heat or water.
- Overloading sockets with too many adaptors and plugs.

Most electrical appliances are connected to the mains supply using cable and a three-pin plug.

The outer cover of a three-pin plug is made of plastic or rubber, as these are good insulators. The pins of the plug are made of brass. Brass is a good conductor. It is also hard and will not rust or oxidise.

It is important that the cable grip is fastened tightly over the cable. There should be no bare wires showing inside the plug and the correct cable must be connected firmly to the terminal of the correct pin.

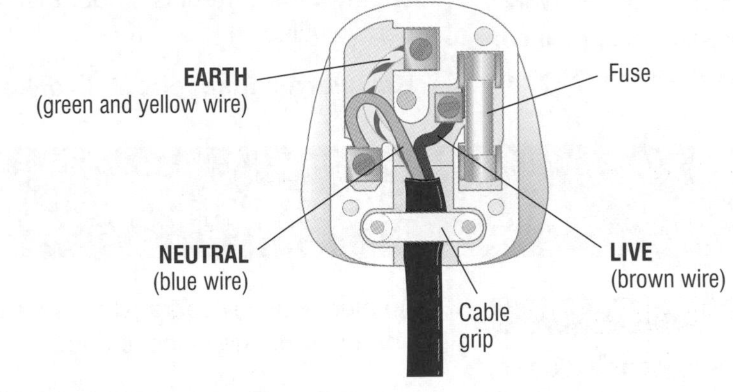

Inside a three-pin plug

- The brown wire is connected to the live pin.
- The blue wire is connected to the neutral pin.
- The green-yellow wire (of a three-core cable) is connected to the earth pin. A two-core cable does not have an earth wire.

Appliances with metal cases must be earthed – the case is attached to the earth wire in the cable. Appliances with plastic cases do not need to be earthed. They are said to be double insulated and are connected to the supply with two-core cable containing just a live and a neutral wire.

Key words: insulator, three-pin plug, cable, live, neutral, earth

CHECK YOURSELF

1 Why must appliances with metal cases be earthed?

2 What colour is the neutral wire?

3 Why are the pins of the plug made of brass?

students' book page 252 P2 6.3 Fuses

KEY POINTS

1. A fuse contains a thin wire that heats up and melts and cuts the current off if too much current passes through it.
2. A circuit breaker is an electromagnetic switch that opens (i.e. 'trips') and cuts the current off if too much current passes through it.

EXAM HINTS

Make sure you can explain how the earth wire and the fuse work together to protect an appliance.

Appliances with metal cases need to be earthed. Otherwise if a fault develops and the live wire touches the metal case the case becomes live and could give a shock to anyone that touches it.

If a fault develops in an earthed appliance a large current flows to earth and melts the fuse, disconnecting the supply.

A fuse must be put in the live wire so that if it melts it cuts off the current. The rating of the fuse should be slightly higher than the normal working current of the appliance. If it is much higher it will not melt soon enough. If it is not higher than the normal current it will melt as soon as the appliance is switched on.

A circuit breaker can be used in place of a fuse. This is an electromagnetic switch that opens and cuts off the supply if the current is bigger than a certain value.

Key words: fuse, circuit breaker

CHECK YOURSELF

1. What is a circuit breaker?
2. What happens if a fault develops in an earthed appliance?
3. Why do appliances with plastic cases not need to be earthed?

students' book page 254 P2 6.4 Electrical power and potential difference

KEY POINTS

1. The power supplied to a device is the energy transfered to it each second.
2. Electrical power supplied (watts, W) = current (amperes, A) × potential difference (volts, V)

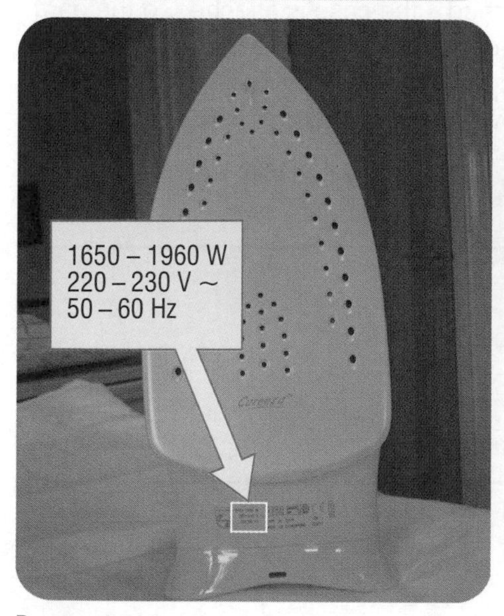

Power rating

An electrical device transforms electrical energy to other forms and transfers energy from one place to another.

The rate at which it does this is called the power. Power can be calculated using:

$$\text{power (watts, W)} = \frac{\text{energy transformed (joules, J)}}{\text{time (seconds, s)}}$$

In an electric circuit it is more usual to measure the current through a device and the potential difference across it.

We can also use current and p.d. to calculate the power of a device:

$$\text{power (watts, W)} = \text{current (amperes, A)} \times \text{potential difference (volts, V)}$$

Electrical appliances have their power rating shown on them. The p.d. of the mains supply is 230 V. So this equation can be used to calculate the normal current through an appliance and so work out the size of fuse to use.

Key words: power, energy

GET IT RIGHT!

Practice calculating the current in appliances of different powers so you can choose the correct fuse.

CHECK YOURSELF

1. What is the power of a mains appliance that takes a current of 10 A?
2. How much energy is transformed when a 3000 W appliance is used for 30 minutes?
3. What fuse should be used in a 500 W mains heater?

students' book page 256

P2 6.5 Electrical energy and charge

KEY POINTS

1 An electric current is the rate of flow of charge.
2 When charge flows through a resistor, electrical energy is transformed into heat energy.

GET IT RIGHT!

When charge flows through a resistor electrical energy is converted to heat. Since almost every component has electrical resistance, including connecting wires, when a charge flows in the circuit the components will heat up. This means that most electrical appliances have vents to keep them cool.

When charge flows through an appliance, electrical energy is transformed to other forms. In a resistor electrical energy is transformed into heat.

HIGHER

The amount of energy transformed can be calculated using the equation:

energy transformed = potential difference × charge
(joules, J) (volts, V) (coulombs, C)

When there is a current of one amp for one second the charge flowing is one coulomb. The equation relating charge, current and time is:

charge = current × time
(coulombs, C) (ampere, A) (second, s)

Key words: energy, current, charge

Energy transformations in a circuit

CHECK YOURSELF

1 What is the unit of charge? [Higher Tier only]
2 How much energy is transformed to heat when a charge of 200 C flows through a heater that has a potential difference across it of 230 V? [Higher Tier only]
3 How much charge flows past a particular point in a circuit when a current of 2 A flows for 2 minutes? [Higher Tier only]

P2 6 End of chapter questions

1 **The UK mains supply has a frequency of 50 Hz. What does this mean?**

2 **What is the difference between d.c. and a.c.?**

3 **How should the cable grip on a plug be fixed?**

4 **What does the cover on an earth wire look like?**

5 **In a three-pin plug, which terminal is the fuse connected to?**

6 **Why must the fuse used in an appliance have a slightly higher rating than the normal working current?**

7 **What is the equation that relates power, current and potential difference?**

8 **What is the current through a 2300 W mains heater?**

9 **What is the equation that relates energy transformed, potential difference and charge? [Higher Tier only]**

10 **What is measured in coulombs? [Higher Tier only]**

P2 7 Pre Test: Nuclear physics

1. What is the atomic number of a nucleus?
2. What is the mass number of a nucleus?
3. What was the plum pudding model of the atom? [Higher Tier only]
4. What material were alpha particles fired at in Rutherford's experiment? [Higher Tier only]
5. What is nuclear fission?
6. Most uranium is uranium-238. What does the 238 tell you about the structure of the nucleus?
7. What is nuclear fusion?
8. Why do nuclei repel each other?

students' book page 262

P2 7.1 Nuclear reactions

KEY POINTS

		Change in the nucleus	Particle emitted
1	**α decay**	The nucleus loses 2 protons and 2 neutrons	2 protons and 2 neutrons emitted as an α particle
2	**β decay**	A neutron in the nucleus changes into a proton	An electron is created in the nucleus and instantly emitted

GET IT RIGHT!

When a nucleus emits gamma radiation there is no change in the atomic number or the mass number, because a gamma ray is an electromagnetic wave which has no charge and no mass.

An atom has a nucleus, made up of protons and neutrons, surrounded by electrons.

The table below gives the relative masses and the relative electric charges of a proton, a neutron and an electron:

	Relative mass	Relative charge
proton	1	+1
neutron	1	0
electron	0.0005	−1

In an atom the number of protons is equal to the number of electrons, so the atom has no overall charge. If an atom loses or gains electrons it becomes charged and is called an 'ion'.

All atoms of a particular element have the same number of protons. Atoms of the same element that have different numbers of neutrons are called 'isotopes'.

The number of protons in an atom is called its 'atomic number'.

The total number of protons and neutrons in an atom is called its 'mass number'.

An alpha particle consists of two protons and two neutrons. When a nucleus emits an alpha particle the atomic number goes down by two and the mass number goes down by four.

For example, radium emits an alpha particle and becomes radon.

$$^{226}_{88}Ra \rightarrow ^{222}_{86}Rn + ^{4}_{2}\alpha$$

A beta particle is a high speed electron from the nucleus. It is emitted when a neutron in the nucleus changes to a proton and an electron. The proton stays in the nucleus so the atomic number goes up by one and the mass number is unchanged. The electron is instantly emitted.

For example, carbon-14 emits a beta particle when it becomes nitrogen:

$$^{14}_{6}C \rightarrow ^{14}_{7}N + ^{0}_{-1}\beta$$

Background radiation is the radiation that is around us all the time. It comes from many different sources such as cosmic rays, from rocks, or from nuclear power stations.

Key words: atomic number, mass number

EXAM HINTS

Make sure you can use nuclear equations to show how the atomic number and mass number change when alpha or beta particles are emitted.

CHECK YOURSELF

1 Where does background radiation come from?

2 What happens to the mass number of a nucleus when it emits a beta particle?

3 What happens to the atomic number of a nucleus when it emits an alpha particle?

students' book page 264

P2 7.2 The discovery of the nucleus

HIGHER

KEY POINTS

1 Alpha particles in a beam are sometimes scattered through large angles when they are directed at a thin metal foil.

2 Rutherford used the measurements from alpha-scattering experiments to prove that an atom has a small positively charged central nucleus where most of the mass of the atom is located.

At one time scientists thought that atoms consisted of spheres of positive charge with electrons stuck into them, like plums in a pudding. So this became known as the 'plum pudding' model of the atom.

Rutherford, Geiger and Marsden devised an alpha particle scattering experiment, in which they fired alpha particles at thin gold foil.

Most of the alpha particles passed straight through the foil. This means that most of the atom is just empty space.

Some of the alpha particles were deflected through small angles, this suggests that the nucleus has a positive charge. A few rebound through very large angles. This suggests that the nucleus has a large mass and a very large positive charge.

Key words: alpha particle scattering

GET IT RIGHT!

The alpha particle has a positive charge. Because some of the alpha particles rebound, they must be repelled by another positive charge.

CHECK YOURSELF

1 What was Rutherford's alpha particle scattering experiment?

2 Why did most alpha particles pass straight through the foil in Rutherford's experiment?

3 What did the alpha particle scattering experiment suggest about the structure of the nucleus?

students' book page 266

P2 7.3 Nuclear fission

KEY POINTS

1 Nuclear fission occurs when a neutron collides with and splits a uranium-235 nucleus or a plutonium-239 nucleus.
2 A chain reaction occurs when neutrons from the fission go on to cause further fission.
3 In a nuclear reactor one fission neutron per fission on average goes on to produce further fission.

BUMP UP YOUR GRADE

Make sure that you can draw a simple diagram to show a chain reaction.

Nuclear fission is the splitting of an atomic nucleus.

There are two fissionable isotopes in common use in nuclear reactors, uranium-235 and plutonium-239.

Naturally occurring uranium is mostly uranium-238, which is non-fissionable. Most nuclear reactors use 'enriched' uranium that contains 2–3% uranium-235.

For fission to occur the uranium-235 or plutonium-239 nucleus must absorb a neutron. The nucleus then splits into two smaller nuclei and 2 or 3 neutrons and energy is released. The energy released in such a nuclear process is much greater than the energy released in a chemical process such as burning.

The neutrons produced go on to produce further fissions, starting a chain reaction.

In a nuclear reactor the process is controlled, so one fission neutron per fission on average goes on to produce further fission.

Key words: fission, fissionable isotope, chain reaction

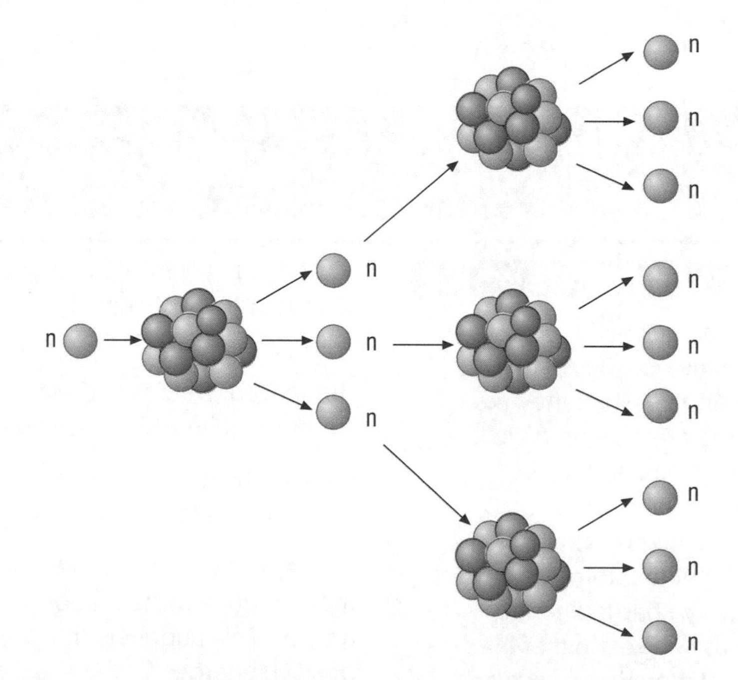

A chain reaction

CHECK YOURSELF

1 What is 'enriched' uranium?

2 What happens for fission to occur?

3 Which two fissionable isotopes are used in nuclear reactors?

students' book page 268

P2 7.4 Nuclear fusion

KEY POINTS

1 Nuclear fusion occurs when two nuclei are forced close enough together so they form a single larger nucleus.
2 Energy is released when two light nuclei are fused together.
3 A fusion reactor needs to be at a very high temperature before nuclear fusion can take place.

AQA EXAMINER SAYS...

In an examination, students often confuse fission and fusion. Make sure that you can explain the difference between them.

Nuclear fusion is the joining of two atomic nuclei to form a single, larger nucleus.

During the process of fusion energy is released. Fusion is the process in which energy is released in stars.

There are enormous problems with producing energy from nuclear fusion in reactors. Nuclei approaching each other will repel one another due to their positive charge. To overcome this the nuclei must be heated to very high temperatures to give them enough energy to overcome the repulsion and fuse. Because of the enormously high temperatures involved the reaction cannot take place in a normal 'container', but has to be contained by a magnetic field.

Key words: fusion

A fusion reaction

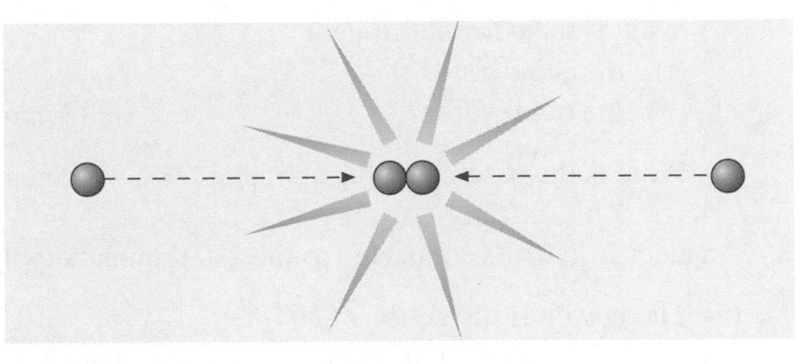

CHECK YOURSELF

1 By what process is energy released in stars?
2 How can nuclei be made to come close enough to fuse?
3 How are nuclei contained in a fusion reactor?

P2 7 End of chapter questions

1 **What happens to the mass number of a nucleus when it emits an alpha particle?**
2 **What happens to the atomic number of a nucleus when it emits a beta particle?**
3 **What particles were used in Rutherford's scattering experiment? [Higher Tier only]**
4 **What does most of an atom consist of?**
5 **What is a 'fissionable isotope'?**
6 **What is a 'chain reaction'?**
7 **Where does most nuclear fusion occur?**
8 **Why do nuclei not normally become close enough to fuse?**

EXAMINATION-STYLE QUESTIONS

1 The diagram shows a parallel circuit.

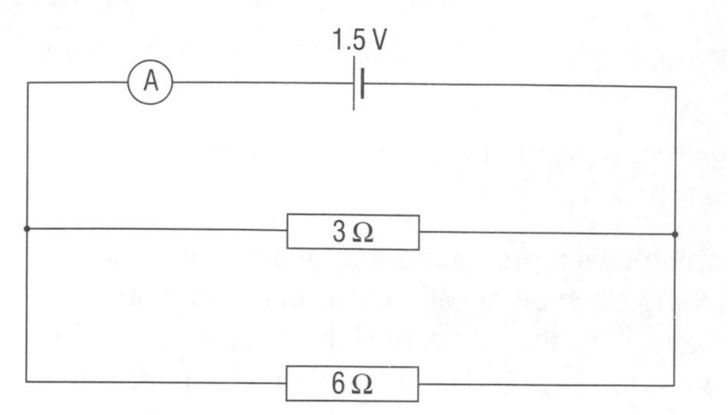

(a) What is the potential difference across:
  (i) the 3 Ω resistor
  (ii) the 6 Ω resistor? (2 marks)

(b) What is the current through:
  (i) the 3 Ω resistor
  (ii) the 6 Ω resistor? (3 marks)

(c) What is the reading on the ammeter? (2 marks)

2 An electric kettle is connected to the 230 V mains supply.

(a) The power of the kettle is 2300 W. What is the current through the kettle? (3 marks)

(b) 3 A, 10 A and 13 A fuses are available. What size fuse should be used in the plug for the kettle? (1 mark)

(c) When the kettle is used to boil a litre of water, 2400 C of charge flows through the kettle. How long, in minutes, does it take to boil a litre of water? (4 marks) [Higher]

3 The diagram represents three different atoms X, Y and Z.

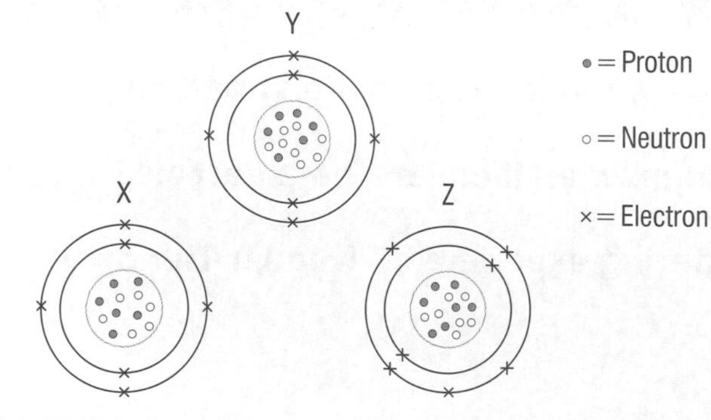

(a) Which two of the atoms have the same mass number? (1 mark)

(b) Which two of the atoms have the same atomic number? (1 mark)

(c) Which two of the atoms are from the same element? (1 mark)

(d) What must happen to atom Z to turn it into a positively charged ion? (1 mark)

(e) Atom Y decays by the emission of a beta particle. Explain what happens to the numbers of protons and neutrons in its nucleus. (2 marks)

4 A student is investigating how the potential difference across a resistor varies with the current through it. The student connects a cell, a resistor and a variable resistor in series.

(a) Draw a circuit diagram for the circuit. (4 marks)

(b) (i) How should the student connect an ammeter in the circuit to measure the current in the resistor?
  (ii) How should the student connect a voltmeter in the circuit to measure the potential difference across the resistor? (2 marks)

(c) Sketch a graph to show how the potential difference across the resistor (x-axis) varies with the current through it (y-axis). (2 marks)

5 The process of nuclear fission may lead to a chain reaction.

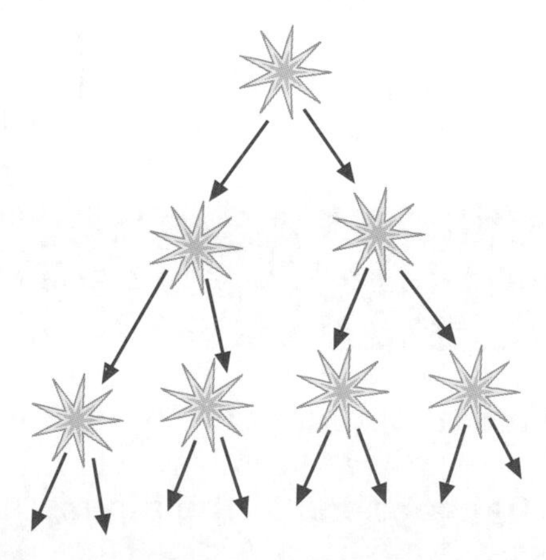

Explain the process of nuclear fission and how it can lead to a chain reaction. (6 marks)

There are 8 possible marks in total for this Higher Tier question.

This answer is worth 6 marks.

The responses worth a mark are underlined in red.

We can improve the answer in several ways:

This does not score a mark. A vacuum means that there are no gas molecules to get in the way and change the path of the alpha particles.

Stating 'this means that most of the atom is empty space' would gain the second mark.

The Rutherford and Marsden scattering experiment led to the 'plum pudding' model of the atom being replaced by the nuclear model.

(a) What is the 'plum pudding' model of the atom? *(2 marks)*

(b) The diagram shows the apparatus used for the scattering experiment to investigate the structure of the atom.

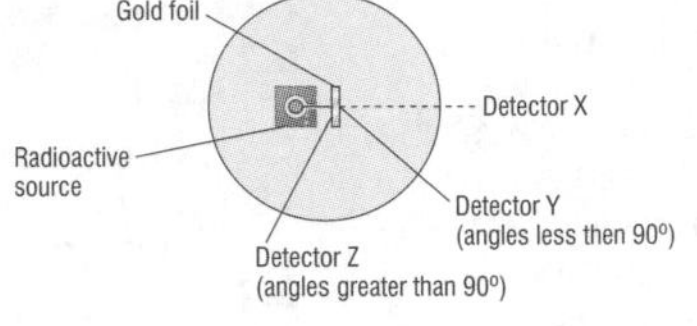

(i) What is the charge on the particles emitted by the radioactive source? *(1 mark)*

(ii) The experiment took place in a vacuum. Suggest why. *(1 mark)*

(iii) Explain why detector X detected the most particles. *(2 marks)*

(iv) Some particles were scattered through very large angles and were detected by detector Z.
What does this suggest about the structure of a gold nucleus? *(2 marks)*

(a) The plum pudding model says that the atom is a blob of positive matter with electrons stuck in like plums in a pudding.

(b) (i) Alpha particles are positively charged.

(ii) Easier to detect the particles.

(iii) Most particles go straight through and are not deflected.

(iv) The nucleus has a large positive charge and a large mass.

There are 8 marks available for this question.

This answer is worth 6 marks.

The responses worth a mark are underlined in red.

We can improve the answer in several ways:

The formula is correct and gets a mark. However, there are two cells so the supply p.d. of 1.5V put into the equation is wrong. This makes the final answer wrong so the last 2 marks are lost.

The incorrect value from part (c) is put into the equation, but everything else is correct, so both marks are gained. Also, the working is shown at each stage.

The diagram below shows a circuit with a d.c. supply.

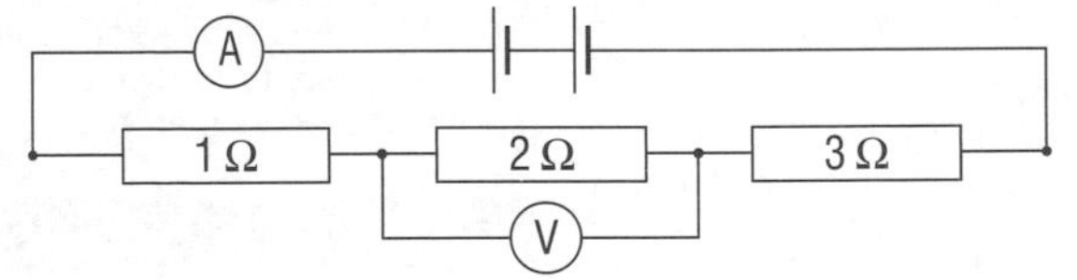

The potential difference of each cell is 1.5 V.

(a) What is the potential difference supplied by the battery? *(1 mark)*

(b) What is the total resistance in the circuit? *(2 marks)*

(c) Calculate the reading on the ammeter. *(3 marks)*

(d) Calculate the reading on the voltmeter. *(2 marks)*

(a) 1.5V + 1.5V = 3V

(b) Total resistance = 1Ω + 2Ω + 3Ω
Total resistance = 6Ω

(c) Current = supply p.d./total resistance
Current = 1.5V/6Ω
Current = 0.25 A

(d) p.d. = current × resistance
p.d. = 0.25 A × 2Ω
p.d. = 0.5V

B2 Answers to questions

Chapter 1

Pre Test

1 It controls the activities of the cell.
2 The cell membrane.
3 The cytoplasm.
4 To produce proteins (protein synthesis).
5 Chloroplasts (containing chlorophyll).
6 Mitochondria.
7 There are many different functions (jobs) that plant and animal cells have to carry out.
8 The random movement of particles from an area of high concentration to an area of lower concentration.
9 Many examples, e.g. oxygen, carbon dioxide, glucose.
10 It takes place through a partially permeable membrane and involves only the movement of water.

Check yourself

1.1
1 Three of: cell membrane, nucleus, cytoplasm, mitochondria, ribosomes.
2 In the mitochondria.
3 A number of examples including carbon dioxide, oxygen and glucose.

1.2
1 Egg cells, muscle cells, white blood cells (many others).
2 It is long and has many endings to communicate with other nerves or effectors.
3 Differentiation.

1.3
1 They naturally vibrate as they have energy.
2 The diffusion of glucose into respiring cells.
3 Cells need a constant supply of materials and to get rid of waste products.

1.4
1 It involves the random movement of particles.
2 A membrane that only allows the smaller molecules (particles) through.
3 For support, and chemical reactions take place in solution.

End of chapter questions

1 To synthesise (make) proteins.
2 It is streamlined for swimming and has lots of mitochondria to produce energy.
3 Particles naturally vibrate – the cell does not need to make them move.
4 It also passes from the stronger to the weaker solution, as the movement of the water molecules is random. However the net movement is to the stronger solution.
5 Chlorophyll.
6 It is for support.
7 It is at a higher concentration in the blood than in the cell.
8 The cell membrane.

Chapter 2

Pre Test

1 The Sun.
2 Glucose.
3 Carbon dioxide.
4 The enzymes do not work so effectively and the molecules move more slowly.
5 There is more light so more energy for the process, and more heat so the molecules move faster colliding more often and with more energy.
6 The temperature will limit the rate of photosynthesis anyway, so increasing the amount of light would have no effect.
7 In dense vegetation, where photosynthesis is taking place rapidly and temperature is high, e.g. in a tropical rainforest.
8 To make their own proteins for growth.
9 Stunted.
10 They cannot make green chlorophyll.

Check yourself

2.1
1 The Sun.
2 Carbon dioxide and water.
3 Oxygen is needed by plants and animals for respiration – the release of energy from food.

2.2
1 The molecules move more slowly, so collisions are less forceful and occur less often.
2 It will be hot enough anyway so you will be wasting energy. It may even become so hot that enzymes denature.
3 In a rapidly photosynthesising area where it is warm, e.g. in a tropical rainforest.

2.3
1 So that the substance does not affect osmosis (as it does not dissolve and alter the strength of a solution).
2 It is combined with other nutrients to form new materials.
3 The release of energy from food.

2.4
1 Through the roots.
2 To make chlorophyll for photosynthesis.
3 It means a 'lack' of something.

End of chapter questions

1 It is trapped by chlorophyll.
2 Temperature.
3 So that it has no effect on osmosis.
4 It will be stunted (shorter).
5 Glucose (sugar).
6 At the bottom of hedgerows, smaller plants in a field of tall growing crops, behind a building, behind a wall – anywhere there is some shade.
7 They cannot make enough protein, protein is necessary for growth.
8 It is stored as starch.

Chapter 3

Pre Test

1 A picture showing the mass of living material at each stage of a food chain.
2 One organism, e.g. a bush, can support large numbers of other organisms so the pyramid can have a very unusual shape. However the mass of the tree is much greater than the mass of the organisms it supports.
3 A number of reasons including: not all the food is digested, the animal needs to keep warm, the animal will move around using up energy.
4 Birds need to maintain a constant temperature. The colder it is, the more energy they need to keep warm.
5 It is expensive as a lot of energy is lost when animals eat plants – not all of the plant is converted into meat (perhaps 20%).
6 By keeping the number of stages in the food chain as few as possible.
7 Decay or decomposition.
8 Microorganisms (bacteria and fungi).
9 Photosynthesis.
10 All of the nutrients are recycled and are re-used by organisms.

Check yourself

3.1
1 The mass of living material.
2 A tree can support many caterpillars, which support a few birds.
3 Each stage can be drawn in proportion to the other stages depending on the mass of organisms.

3.2
1 The energy at each stage that is not converted into energy in the next stage.
2 Because to keep warm, as you lose heat to the surroundings, requires a lot of energy.
3 Energy is required by muscles to contract and cause movement.

3.3
1 Less energy wastage between the stages of the food chain.
2 It will lose a lot of heat to the surroundings and will need energy to replace the energy lost.
3 They prevent the animal from any movement so that it uses less energy.

3.4
1 An organism that feeds on waste.
2 Enzymes work faster in warmer conditions.
3 Decomposition or break down of material.
4 Bacteria or fungi.

3.5
1 Respiration.
2 Detritus feeders or decay organisms (microorganisms).
3 Returning it to the atmosphere so that it is not all 'locked' up by organisms.

End of chapter questions

1 Respiration.
2 It costs too much in energy losses from plants to produce it. There would not be enough to eat.
3 The enzymes work more quickly.
4 Plants remove carbon from the atmosphere in photosynthesis. Plants return carbon to the atmosphere in respiration.
5 Biomass shows the amount of living material at each stage. With a pyramid one tree may support several thousand insects, so the pyramid does not accurately reflect what is happening.
6 It passes out of an animal as faeces and enters the decay process.
7 Detritus feeders.
8 They respire 24 hours per day, all of the time.

EXAMINATION-STYLE QUESTIONS

1 (a) A: cell membrane.
B: nucleus.
C: cytoplasm. (3 marks)
(b) Long – impulses travel long distances.
Lots of connections – to other nerves. (2 marks)
(c) (i) Any two of: vacuole, chloroplast, cell wall. (2 marks)
(ii) A vacuole contains the cell sap.
A chloroplast contains chlorophyll / traps energy.
A cell wall – adds strength to the cell. (2 marks)

2 (a) From the top down:
Large bird (sparrowhawk), small bird (blue tit), caterpillar, tree. (3 marks)
(b) Any three of: Some is not digested.
Some is excreted.
Some is used to keep warm.
Some is used for movement. (3 marks)

(c) Keep the animals at a warmer temperature.
Stop the animals from moving around. (2 marks)

3 (a) (i) Respiration / energy. (1 mark)
All of the time / 24 hours per day. (1 mark)
(ii) To produce amino acids or to produce proteins.
For growth. (2 marks)
(iii) Small / stunted. (1 mark)
(b) The leaves are yellow, as they cannot produce chlorophyll. (2 marks)
(c) Warm and wet. (2 marks)
(d) All of the materials are recycled within the community. (1 mark)

4 (a) (i) The (random) movement of particles, from an area of high concentration, to an area of lower concentration. (3 marks)
(ii) A: carbon dioxide.
B: oxygen. (2 marks)
(b) The particles are moving (randomly) anyway.
The energy source is the warmth / temperature / Sun. (2 marks)
(c) (i) Any two of: Movement of particles (random), from higher to lower concentration, does not need energy from the cell. (2 marks)
(ii) It only involves water molecules.
It takes place through a partially permeable membrane. (2 marks)

Chapter 4

Pre Test

1 A catalyst that works in a living organism – an enzyme.
2 Respiration and photosynthesis (though there are many more).
3 It is where the reacting molecules (particles) fit.
4 In the mitochondria.
5 Proteins.
6 A gland.
7 Lipases.
8 In the liver.
9 In the stomach.
10 It is sweeter than glucose so you need less to sweeten foods. It is therefore useful in 'slimming' foods.

Check yourself

4.1
1 Proteins.
2 The area of the enzyme that the reacting particles fit into.
3 Speeding up a chemical reaction.

4.2
1 Enzymes work more effectively, molecules move faster so have more collisions and collide with more force.
2 Their shape changes, they denature. The reacting molecules will not fit into the active site.
3 Enzymes work best in a narrow pH range. If it varies too much then they might denature.

4.3
1 In the mitochondria.
2 Energy (from respiration). Muscles also need a stimulus.
3 Proteins.

4.4
1 Fats and oils.
2 Amylase, protease and lipase (all of them!).
3 Fatty acids and glycerol.

4.5
1 (Slightly) alkali.
2 To neutralise stomach acid.
3 Hydrochloric acid.

4.6
1 You do not need very much as it is much sweeter than glucose.
2 It makes digestion of the food easier for the baby, as some has been partly digested already.
3 It would denature the enzymes so the washing powder would no longer work.

End of chapter questions

1 The amount of energy necessary to start a reaction.
2 The area of the enzyme where the reacting substances fit in.
3 To build larger molecules from smaller ones, muscle contraction and to maintain a constant body temperature (mammals and birds only).
4 The breakdown of proteins into amino acids by the enzyme protease.
5 We use the term 'denatured'.
6 Carbon dioxide and water.
7 The pancreas, stomach and small intestine.
8 It is made by the liver and stored in the gall bladder.

Chapter 5

Pre Test

1 Carbon dioxide and water.
2 They are converted into urea and excreted.
3 In the liver.
4 The ion content of cells is linked to osmosis so a cell could take up or lose too much water if it was not controlled.
5 The thermoregulatory centre of the brain and receptors in the skin.
6 The energy needed for the sweat to evaporate comes from the surface of the skin.
7 In the sweat glands.
8 It increases the respiration rate. Some energy produced is released as heat.
9 Insulin.
10 In the pancreas.

Check yourself

5.1
1 Respiration.
2 Urea.
3 In the bladder.

5.2
1 The thermoregulatory centre of the brain and receptors in the skin.
2 In sweat glands under the surface of the skin.
3 Some of the energy released from the increased respiration is released as heat.

5.3
1 The pancreas.
2 In the liver.
3 Diabetes.

End of chapter questions

1 It affects the strength of the solution in the cell and so affects osmosis.
2 To monitor the body's temperature and then coordinate a response if the core body temperature is too high or too low.
3 The sweat evaporates from the skin's surface. The energy for this comes from the skin.
4 The pancreas detects a high sugar level, the pancreas produces insulin, some glucose is converted to glycogen and stored in the liver.
5 By radiation.
6 Through the kidneys, dissolved in water to form the urine.
7 The blood vessels constrict.
8 Osmosis.

Chapter 6

Pre Test

1 In all body cells (but not sex cells).
2 Two cells are formed.
3 For growth and to replace cells.
4 Four cells.
5 Alleles.
6 They can still differentiate into new specialised cells and so are seen as a possible cure for some diseases and conditions.
7 23 pairs.
8 Deoxyribonucleic acid.
9 An allele that 'masks' the effect of the other allele (the recessive one).
10 The nervous system.

Check yourself

6.1
1 To produce new cells for growth and to replace cells.
2 They have exactly the same genetic make up as the parent cell.
3 Two cells.

6.2
1 They can differentiate and grow into more specialised cells – this may help cure some conditions in the future.
2 Stem cells that are able to differentiate into many different specialised cells are found in developing embryos. These must be destroyed to get the stem cells.
3 Adult bone marrow contains stem cells that will only develop into a limited range of specialised cells.

6.3
1 Four sex cells.
2 Twice.
3 There is a mixing of genetic information. Half of the information comes from the male parent and the other half from the female parent.

6.4
1 Gregor Mendel.
2 Deoxyribonucleic acid.
3 Short lengths of DNA.

6.5
1 23 pairs (46).
2 XY.
3 An allele where the effect is masked by a dominant allele. An organism needs both recessive alleles if it is to show that characteristic.

6.6
1 The nervous system.
2 An organism that carries an allele (recessive) for the condition but does not have the symptoms.
3 They are looking for alleles that will result in the organism having some form of disease or disorder.

End of chapter questions

1 The stem cells develop into specialised cells.
2 Four sex cells that are all different to each other and the parent cell.
3 An allele that masks the effect of the recessive allele. An organism needs only one of these alleles in the pair to show that characteristic.
4 A short length of DNA that controls one characteristic.
5 They make a copy of themselves.
6 23 pairs.
7 Pairs of genes controlling the same characteristic.
8 A disorder of cell membranes.

EXAMINATION-STYLE QUESTIONS

1 (a) (i) The lungs.
(ii) The liver.
(iii) The kidney.
(iv) The skin. (4 marks)
(b) As it evaporates, it takes energy from the skin, cooling the skin down. (3 marks)
(c) Any two of: muscles contract, they respire to produce the energy necessary, some energy is lost as heat. (2 marks)

2 (a) In developing embryos and adult bone marrow. (2 marks)

(b) Any two of: stem cells can differentiate (develop), into different types of cell, possible cure for certain conditions / diseases. (2 marks)

(c) The most useful stem cells come from developing embryos, you destroy the embryo by taking the stem cells, some people argue, therefore, that you have destroyed a 'life'. (3 marks)

3 (a) (i) 2 (1 mark)

(ii) By carrying out the readings more than once to avoid anomalous results (it does state all other variables were kept the same). (1 mark)

(iii) By using a narrower band of pH readings with smaller intervals between them, e.g. 1.2, 1.4, 1.6 etc. (2 marks)

(b) In the stomach, the only part where protease works in acid conditions. (2 marks)

(c) (i) Lipase digest fats, into fatty acids, and glycerol. Amylase digests starch, into glucose (sugar). (5 marks)

(ii) Two of: In the mouth (salivary glands), pancreas, small intestine. (2 marks)

(d) To neutralise, acid from the stomach. (2 marks)

4 (a) It is the energy source for cells. (1 mark)

(b) The pancreas. (1 mark)

(c) The pancreas produces insulin. (1 mark) Two of: excess glucose is converted into glycogen, by the liver, glycogen is stored in the liver (and some muscles). (2 marks)

(d) Either through eating an appropriate diet, or through insulin injections. (2 marks)

5 (a) Diagram should show that mother is Hh and father is hh. D has inherited the dominant H allele from his mother and the recessive h allele from his father, and therefore has the disease. (Note that mother cannot be HH or all of her children would have the disease.) (3 marks)

(b) Inherited the recessive gene from the father, and the recessive gene from the mother. (2 marks)

(c) 50%
The recessive from parent F has a 1 in 2 chance of pairing with the dominant or the recessive gene from the person with H. (2 marks)

(d) (i) Two of: It codes for amino acids, these link to form proteins, proteins control the characteristics we have. (2 marks)

(ii) A gene. (1 mark)

C2 Answers to questions

Chapter 1

Pre Test

1 Protons and neutrons in the nucleus with electrons surrounding it.
2 The number of protons in an atom of the element.
3 In energy levels (shells).
4 They have the same number of electrons in their highest energy level (outer shell).
5 They are stable arrangements.
6 They transfer (from metals to non-metals) or they are shared.
7 Strong electrostatic forces between oppositely charged ions.
8 It contains equal numbers of sodium and chloride ions / the ratio of sodium ions to chloride ions is 1:1.
9 A pair of electrons shared by two atoms.
10 Either H–O–H or dot/cross diagram with a pair of dot/crosses between each of 2 Hs and an O and 4 more dot/crosses giving a total of 8 around the O.
11 In regular (giant) structures / lattices.
12 By metallic bonds – strong electrostatic forces between delocalised electrons and positive metal ions.

Check yourself

1.1
1 Protons, neutrons and electrons.
2 Proton +1, neutron 0, electron –1.
3 12

1.2
1 They represent electron energy levels.
2 Two concentric circles with 2 dots or crosses on the inner circle and 4 dots or crosses on the outer circle (and a dot at the centre for the nucleus).
3 Lithium 2,1; nitrogen 2,5; magnesium 2,8,2.

1.3
1 They have stable electronic arrangements/structures (*not* full outer shells).
2 K^+ 2,8,8; Mg^{2+} 2,8; O^{2-} 2,8.
3 Lithium atom: inner circle with 2 dots or crosses, outer circle with 1 dot or cross.
Fluorine atom: inner circle with 2 dots/crosses, outer circle with 7 dots/crosses.
Lithium ion: one circle with 2 dots/ crosses surrounded by square bracket with + sign outside top right of bracket.
Fluoride ion: as fluorine atom but with 8 dots/crosses in outer circle and square bracket with – sign outside top right.

1.4
1 Strong electrostatic forces between oppositely charged ions.
2 Equal numbers of sodium and chloride ions / ratio of 1 sodium ion : 1 chloride ion (no mention of molecules).
3 CaF_2

1.5
1 A shared pair of electrons (between two atoms).
2 4
3 N at centre with 3 separate Hs around it, with a dot and a cross between each H and the central N, and 2 more dots/crosses at a similar distance from the N (total 5 dots + 3 crosses *or* 3 dots + 5 crosses).

1.6
1 In regular patterns / lattices / giant structures.
2 On surface of some metals / when metals are displaced from solutions.
3 Electrons from highest energy level (outer electrons) that move freely within a metal (structure) or from atom to atom in a metal.
4 Electrostatic forces (between electrons and positive ions).

End of chapter questions

1 Both protons and electrons have the same charge and atoms are neutral.
2 9 protons, 9 electrons.
3 2,8,8,1.
4 All have one electron in their outer shell.
5 An unreactive gas, with a stable arrangement of electrons, in Group 0 of the periodic table.
6 The sodium atom loses its outer electron to form an ion with a single positive charge. The fluorine atom gains an electron to form an ion with a single negative charge. The sodium electron has transferred to the fluorine atom.
7 Strong electrostatic forces between oppositely charged ions (ionic bonds).
8 It has equal numbers of potassium ions and chloride ions (in the lattice).
9 The number of bonds formed = 8 – the Group number (8 minus the Group no.).
10 H–S–H is the simplest way to show this. Could also be shown as a dot/cross diagram with S and 2 Hs with a dot and cross between each of them, and another 4 dots or crosses around the S.
11 Close together in regular patterns / in layers / in a lattice.
12 By electrostatic forces between the delocalised (free) electrons and positive metal ions.

Chapter 2

Pre Test

1 They have giant structures with strong forces holding the ions together.
2 When they are molten or in solution.
3 There are only weak forces between the molecules so they melt and boil at low temperatures.
4 Molecules do not have a charge.
5 Their atoms can form several strong bonds with each other and/or other atoms.
6 They have very different structures: diamond is a giant 3-D structure, while graphite has giant 2-D layers.
7 The atoms are in layers that slide over each other into new positions.
8 Yes, all metals conduct heat and electricity.
9 The study of very small particles containing a few hundred atoms and about a nanometre in size.
10 As sensors, catalysts, coatings and construction materials, and medical uses.

Check yourself

2.1
1 Many strong forces hold the ions together.
2 Water molecules can split up the lattice.
3 The ions are charged and can move (so they carry the electricity).

2.2
1 It is made of small molecules.
2 It is made of molecules and the molecules are not charged.
3 Weak intermolecular forces / weak attractions between molecules.

2.3
1 A giant covalent structure / a giant structure with atoms joined by covalent bonds.
2 They have regular three-dimensional giant structures.
3 It is made of flat (2-dimensional) molecules with weak forces between the layers.
4 It has delocalised electrons (in the layers).

2.4
1 The layers of atoms can slide over each other into new positions (without the structure breaking apart).
2 The delocalised electrons move through the metal.
3 Delocalised electrons carry the energy.

2.5

1 A few hundred (200–500).
2 Their structure (may be different) and their very small size.
3 Three from: sensors, catalysts, coatings, construction materials, drug delivery/release, microprocessors/computers.

End of chapter questions

1 There are many strong forces to overcome to break down the giant structure so the ions can move about.
2 The ions are free to move in the solution.
3 Weak forces between molecules.
4 The molecules stay as molecules but they move apart and escape from the liquid. (The covalent bonds are *not* broken.)
5 Each atom in the giant structure is held by four strong covalent bonds, and many bonds have to be broken to free the atoms from the lattice.
6 They are covalently bonded in giant flat sheets.
7 The atoms are in layers that can slide over each other without breaking apart because the delocalised electrons keep them held together.
8 Only the delocalised electrons move through the metal (and any that leave the metal are replaced by other electrons from the electric current).
9 They are very small in size, have different structures, have very large surface areas, conduct more easily, and can be sensitive to light, heat, pH, electricity and magnetism.
10 The application of nanoparticles to specific uses.

Chapter 3

Pre Test

1 The number of protons plus the number of neutrons in the atom.
2 Isotopes
3 Atoms are too small to weigh so we use relative masses.
4 An amount of a substance, equal to its formula mass in grams, that contains a certain number of particles.
5 From its formula using relative atomic mass and relative formula mass.
6 Empirical formula is the simplest formula, molecular formula shows the number of atoms in a molecule.
7 The relative amounts of the substances in the reaction.
8 5.6 g
9 The amount of product actually made compared with the maximum amount it is possible to make.
10 It measures how much of the starting materials end up as useful product.
11 By its special symbol, and because it goes in both directions so products react to produce reactants.
12 When a reversible reaction is in a closed system and the rates of forward and backward reactions are equal.
13 By the Haber process from nitrogen and hydrogen.
14 It is a reversible reaction (and because of the conditions used).

Check yourself

3.1

1 Mass number.
2 Atoms of the same element with different numbers of neutrons / atoms with the same atomic number but different mass numbers.
3 $^{16}_{8}O$: p = 8, n = 8, e = 8; $^{19}_{9}F$: p = 9, n = 10, e = 9

3.2

1 (a) $H_2 = 2$
(b) $CH_4 = 16$
(c) $MgCl_2 = 95$
2 18 g
3 $^{12}_{6}C$

3.3

1 71% (or 71.4% or 71.43%)
2 CH_2
3 $FeCl_2$

3.4

1 $H_2 + Cl_2 \rightarrow 2HCl$
2 58.5 g
3 12.5 g

3.5

1 Reactions do not always go to completion and some product may be lost in the process.
2 69.8%
3 56%

3.6

1 A reaction that can go in both directions / a reaction in which products react to produce reactants.
2 It can reach equilibrium / rates of forward and backward reactions become equal.
3 By changing the conditions / by opening the system or allowing product to escape.

3.7

1 Nitrogen (air) and hydrogen (natural gas).
2 nitrogen + hydrogen $\rightleftharpoons$ ammonia
$N_2 + 3H_2 \rightleftharpoons 2NH_3$
3 It is reversible / ammonia breaks down into nitrogen and hydrogen.

End of chapter questions

1 Their mass is too small / negligible.
2 13 protons, 13 electrons, 14 neutrons.
3 62
4 44 g
5 75%
6 Fe_2O_3
7 $CH_4 + 2O_2 \rightarrow CO_2 + 2H_2O$
8 4.25 g
9 63%
10 74.3%
11 The reaction can go both ways: ammonium chloride reacts to produce ammonia and hydrogen chloride *and* hydrogen chloride and ammonia react to produce ammonium chloride.
12 A closed system.
13 450°C and 200 atmospheres pressure.
14 They are recycled.

EXAMINATION-STYLE QUESTIONS

1 (a) Three concentric circles with Na or dot at centre, 2 crosses or dots on inner circle, 8 on next and one on outer circle. (All correct = 2 marks; 1 mark for diagram with one error or omission.)
(b) Two concentric circles with Na or dot at centre, 2 crosses or dots on inner circle, 8 on outer circle (all enclosed in brackets), with + at top right side. (All correct = 2 marks; 1 mark for diagram with one error or omission.)
(c) Circles with 2,8 as for part (b), with O or dot at centre, (all enclosed in brackets) with 2– at top right side. (All correct = 2 marks; 1 mark for diagram with one error or omission.)
(d) Giant lattice / giant structure / many ionic bonds to overcome. (1 mark)
Ionic bonding is strong / oppositely charged ions strongly attract. (1 mark)
(e) Sodium ion has single (positive) charge and oxide ion has double (negative) charge, (1 mark)
so lattice/structure/compound contains twice as many sodium ions as oxide ions *or* empirical formula has 2 Na to each O. (1 mark)

2 (a) (i) 1 (ii) 0 (iii) –1 (3 marks)
(b) (i) 17 protons, 18 neutrons, 17 electrons. (3 marks)
(ii) $^{37}_{17}Cl$ has 2 neutrons more than $^{35}_{17}Cl$. (1 mark)
(c) (i) Relative atomic mass is an average value, (1 mark) that depends on the relative amounts of the isotopes. (1 mark)
(ii) Both have the same number of electrons, (1 mark) so they have the same number in the outer shell / are in the same group of the periodic table / are the same element / chemical properties depend on the number of (outer) electrons. (1 mark)

3 (a) Covalent. (1 mark)
(b) N at centre with 8 dots or crosses around it, 3 H arranged around these, so that there is a pair of dots/crosses between each H and the central N.
e.g. $H : \dot{\underset{\cdot\cdot}{N}} : H$
$\quad\quad H$
(All correct = 2 marks; 1 mark for diagram with one error or omission.)
(c) Made of small molecules, (1 mark) with weak forces between the molecules. (1 mark)

4 (a) ammonium chloride $\rightleftharpoons$ ammonia + hydrogen chloride
(Correct names on either side of equation (1 mark); reversible reaction symbol (1 mark).)
(b) (i) 53.5 g (2 marks for correct answer with units, 1 mark for working: e.g. 14 + 4 + 35.5 or answer without g.)
(ii) 26.2% (2 marks for correct answer, accept 26.17%; 1 mark for working: e.g. 14 × 100 / 53.5.)

5 (a) germanium oxide + hydrogen → germanium + water (2 marks)
(1 mark for reactants, 1 mark for products; accept hydrogen oxide for water.)
(b) $GeO_2 + 4HCl \rightarrow GeCl_4 + 2H_2O$ (2 marks)
(1 mark for 4HCl, 1 mark for $2H_2O$)
(c) Ge^{4+}, O^{2-} (2 marks)
(d) Central Ge surrounded by 4 pairs of dots/crosses, surrounded by 4 Cl, so that there is a pair of dots/crosses between each Cl and Ge.
(All correct = 2 marks; 1 mark for diagram with one error or omission.)
(e) (i) Two from:
Shiny (solid), conducts electricity, forms positive ions / ionic compound (with oxygen), oxide is base / reacts with acid to form a salt.
[High melting point is a neutral statement – could apply to both.] (2 marks)
(ii) Two from:
Brittle (solid), has a giant covalent structure, forms a covalent chloride / its chloride has small molecules, only conducts a small amount of electricity. (2 marks)
(f) (i) 62.4% (2 marks)
(1 mark for working: e.g. 73 × 100 / (73 + 32 + 12))
(ii) 80.2% (2 marks)
(1 mark for working: e.g. 73 × 100 / (73 + 2 + 16))
(iii) Two from:
Better atom economy, other product less harmful to environment / CO_2 produces global warming, water does not (or words to that effect), hydrogen better for sustainable development. (2 marks)

Chapter 4

Pre Test

1 How fast the reaction goes / how quickly reactants are used / how quickly products are formed.

2 By measuring the amount of a reactant used in a certain time or by measuring the amount of a product formed in a certain time.
3 They must collide with sufficient energy to react.
4 Changing conditions changes the frequency of collisions (the number of collisions in a given time, e.g. collisions per second).
5 Particles collide more frequently (more collisions per second) *and* with more energy.
6 The rate doubles.
7 It increases.
8 An increase in pressure increases the rate of reactions between gases.
9 A catalyst changes (increases) the rate of a reaction and is left at the end of the reaction.
10 They speed up reactions and reduce energy costs.

Check yourself

4.1

1 The rate of a reaction measures how fast the reaction goes.
2 Amount of reactant used or product formed and the time.
3 By measuring the volume of gas produced over time (in a certain time).

4.2

1 Reactions happen when particles collide with enough energy to react.
2 Increasing temperature, concentration of solutions, pressure of gases, surface area of solids and using a catalyst.
3 Powders have more (greater) surface area than solids.

4.3

1 Both the number of collisions and the energy of the collisions are increased.
2 Increasing the temperature by 10°C.
3 Decreasing the temperature slows down the rate of reactions that spoil food.

4.4

1 Particles collide more frequently (more often).
2 The particles are closer together (concentration is increased) so they collide more frequently.
3 Equal volumes of the same concentration contain the same number of particles of solute.

4.5

1 They lower the activation energy so more collisions result in reactions.
2 They are not used up in the reaction / can be used over and over again.
3 Catalysts often work with only one type of reaction / are specific for a particular reaction.

End of chapter questions

1 $\text{rate} = \dfrac{\text{amount of reactant used}}{\text{time}}$

or $\dfrac{\text{amount of product formed}}{\text{time}}$

2 Measure volume of gas produced against time (accept mass lost against time *or* change in pH with time).
3 Activation energy.
4 Increase temperature, increase concentration of reactants, increase surface area of magnesium (use smaller pieces of magnesium), use a catalyst [*not increase pressure*].
5 Increasing temperature increases the frequency of collisions *and* the energy of the collisions.
6 It will be one quarter of the original rate.
The dissolved particles are closer together
nd so collide more frequently (more times
r second).
particles are closer at higher pressure
lide more often.

9 They lower the activation energy so more collisions result in reaction.
10 They are not used up (they are left at the end of the reaction).

Chapter 5

Pre Test

1 Exothermic.
2 Endothermic.
3 They are equal.
4 Increasing the temperature increases the amount of products (yield) from the endothermic reaction.
5 Those with different numbers of molecules of gases in the reactants compared to the number of molecules of gases in the products.
6 To produce ammonia as economically as possible.

Check yourself

5.1

1 Endothermic.
2 Heat is given out / the surroundings get hotter / temperature (of surroundings) increases.
3 When they dissolve (or react with saliva) endothermic reaction(s) happen.

5.2

1 Reverse reaction is exothermic.
2 50 kJ taken in (absorbed).
3 Decrease / lower it.

5.3

1 It will have a different number of molecules of gases in the reactants compared with the products.
2 High pressure produces more (higher yield of) ammonia, but very high pressures are expensive.
3 Low temperatures produce more (higher yield of) ammonia, but the reaction is too slow at lower temperatures.

End of chapter questions

1 Two from:
Combustion (burning fuels, burning metals) / respiration / neutralisation.
2 Endothermic reaction(s) happen when they mix with (dissolve in) the water *or* they react with water and take in heat / reactions happen that take in heat.
3 Lower / decrease the temperature.
4 95 kJ are taken in / absorbed.
5 No effect (because there are equal numbers of molecules of gases in the reactants and products).
6 Low temperature and high pressure.

Chapter 6

Pre Test

1 They are broken down (split up) into elements.
2 Non-metallic elements.
3 At the negative electrode.
4 A solution of an ionic compound in water in which the metal that formed the compound is more reactive than hydrogen.
5 Hydrogen, chlorine and sodium hydroxide.
6 The products have many important uses.
7 Copper atoms lose electrons / are oxidised.
8 They form a sludge (from which precious metals are extracted).

Check yourself

6.1

1 A molten ionic compound or a solution containing ions.
2 They gain electrons / are reduced.
3 Zinc and chlorine.

6.2

1 They are discharged / gain electrons / are reduced / form copper atoms.
2 They are discharged / lose electrons / are oxidised / form chlorine atoms / form chlorine molecules.
3 All metals form positive ions (so they are attracted to the negative electrode where they gain electrons).

6.3

1 Sodium / Na^+, hydrogen / H^+, chloride / Cl^-, hydroxide / OH^-.
2 Hydrogen ions and chloride ions are discharged, leaving sodium ions and hydroxide ions in the solution.
3 Hydrogen: making margarine, making hydrochloric acid.
Chlorine: sterilising water, making bleach or disinfectants or plastics.
Sodium hydroxide: making paper or soap or bleach, to control pH.

6.4

1 Impurities decrease conductivity / increase resistance of copper.
2 Copper atoms lose electrons / are oxidised to form copper ions.
3 Copper ions from the solution gain electrons at the negative electrode to form copper atoms that are deposited on the electrode.

End of chapter questions

1 They move towards the electrodes / positive ions are attracted to negative electrode and negative ions are attracted to the positive electrode.
2 Chlorine.
3 Magnesium ions gain electrons / are reduced to form magnesium atoms.
4 Bromine (Br_2).
5 Water (is present and) produces hydrogen ions that are discharged/reduced (rather than sodium ions).
6 Chlorine and sodium hydroxide.
7 They are oxidised / form copper ions (Cu^{2+}) that go into the solution.
8 It (is negatively charged so) attracts copper ions and they gain electrons / are reduced to form copper atoms that are deposited on the electrode.
9 $Al^{3+} + 3e^- \rightarrow Al$
10 $2Cl^- \rightarrow Cl_2 + 2e^-$
or
$2Cl^- - 2e^- \rightarrow Cl_2$

Chapter 7

Pre Test

1 (a) hydrogen ions / $H^+(aq)$
(b) hydroxide ions / $OH^-(aq)$
2 It is very alkaline / it contains a large amount of hydroxide ions / has a high concentration of hydroxide ions.
3 Salt and hydrogen.
4 Magnesium oxide or magnesium hydroxide.
5 Potassium sulfate and water.
6 By mixing solutions containing lead ions and chloride ions / by mixing solutions of a soluble lead salt (e.g. lead nitrate) with a soluble chloride (e.g. sodium chloride).

Check yourself

7.1

1 Hydrogen ions.
2 They produce hydroxide ions in the solution that make it alkaline.
3 Universal indicator or wide range indicator (*not* just indicator).

7.2

1 A salt and water.
2 So that all of the acid is used.
3 Zinc chloride.
4 Copper oxide and nitric acid.

7.3

1 An indicator.
2 Ammonium chloride.
3 Copper sulfate + sodium carbonate.

End of chapter questions

1 It is very acidic / it has a large amount (or high concentration) of hydrogen ions.
2 A base that dissolves in water and forms hydroxide ions in the solution.
3 Magnesium sulfate and hydrogen.
4 Copper chloride and water.
5 There is no visible change when acids react with alkalis.
6 Sodium nitrate and water.
7 $H^+(aq) + OH^-(aq) \rightarrow H_2O(l)$ (state symbols not essential)
8 Any named soluble sulfate, e.g. sodium sulfate (but *not* lead sulfate).

EXAMINATION-STYLE QUESTIONS

1 (a) One mark each for:
Suitable axes, correctly labelled.
All points correctly plotted (within half small square).
Smooth line through points.
Omitting 5 minutes / 65 cm^3. (4 marks)
(b) (i) Slope/gradient was steepest at start. (1 mark)
(ii) Two from:
Concentration of reactants highest at start / concentration of reactants decreases with time.
More particles of reactants at the start / reactants get used up over time.
Greater frequency of collisions at start / more collisions per second at start. (2 marks)
(c) Line with steeper slope initially, (1 mark)
levelling off at same volume as other line. (1 mark)

2 (a) So the ions can move / so the ions are free to move. (1 mark)
(b) Two from:
(Positive) sodium ions are attracted to the (negative) electrode.
Sodium ions gain electrons.
Sodium ions are reduced, form sodium atoms. (2 marks)
(c) Two from:
Chloride ions are attracted to the positive electrode.
Chloride ions lose electrons.
Chloride ions are reduced, form chlorine atoms / form chlorine molecules. (2 marks)
(d) So it does not react with chlorine. (1 mark)
(e) (i) Chlorine, sodium hydroxide. (2 marks)
(ii) Two from:
Water produces H^+ ions / both Na^+ and H^+ ions present.
Sodium is more reactive than hydrogen.
Hydrogen ions discharged/reduced in preference to sodium ions. (2 marks)

3 (a) (i) One mark each for:
Suitable axes, correctly labelled.
All points correctly plotted (within half small square).
Smooth lines through points (no daylight showing at points).
Temperatures correctly labelled. (4 marks)
(ii) 70 atmospheres (+/– 2) at 350°C (1 mark)
(iii) Line drawn between the two lines but closer to 500°C (1 mark)
(b) (i) Reversible reaction. (1 mark)
(ii) Yield better at low temperatures.
Reaction too slow at low temperatures / catalyst does not work at low temperatures. (2 marks)
(iii) High pressure gives high yield.
Very high pressures are too expensive / too dangerous / use too much energy / need very strong equipment. (2 marks)

P2 Answers to questions

Chapter 1

Pre Test

1 Metres per second, m/s.
2 Horizontal line.
3 Acceleration $= \dfrac{\text{change in velocity}}{\text{time taken for the change}}$
4 A deceleration.
5 Acceleration.
6 Horizontal line.
7 Slope $= \dfrac{\text{change in } y \text{ value}}{\text{change in } x \text{ value}}$
8 Slope increases.

Check yourself

1.1
1 Speed.
2 Speed $= \dfrac{\text{distance travelled}}{\text{time taken}}$
3 Metre, m.

1.2
1 Velocity is speed in a given direction.
2 Metres per second squared, m/s^2.
3 Slowing down.

1.3
1 A body travelling at a constant velocity.
2 Straight line graph with a negative slope.
3 Distance travelled.

1.4
1 10 m/s.
2 20 m/s.
3 1 m/s^2.

End of chapter questions

1 A body that is stationary.
2 6.25 m/s.
3 If it is changing direction, its velocity is changing so it is accelerating, even if it is travelling at a steady speed.
4 Acceleration.
5 Deceleration.
6 Area under the graph.
7 10 m/s.
8 99 m.

Chapter 2

Pre Test

1 They are equal.
2 Upwards.
3 A single force that has the same effect on a body as all the original forces acting together.
4 It accelerates the object in the direction of the force.
5 Resultant force = mass × acceleration.
6 3 m/s^2.
7 Time between the driver seeing something and responding to it.
8 Poor roads, poor weather, worn tyres, worn brakes.
9 Velocity reached by an object falling through a fluid when the weight is equal to the resistive forces.
10 Weight = mass × gravitational field strength.

Check yourself

2.1
1 N.
2 Downwards.
3 15 N to the left.

2.2
1 Continues moving at a steady speed in a straight line.
2 Newton.
3 When the direction of the force is opposite to the direction the body is moving in.

2.3
1 Mass $= \dfrac{\text{force}}{\text{acceleration}}$
2 Increases.
3 0.004 m/s^2.

2.4
1 Zero.
2 Stopping distance = thinking distance + braking distance.
3 The greater the speed, the greater the stopping distance.

2.5
1 N/kg.
2 Initially the only force acting is the weight downwards.
3 It reaches terminal velocity.

End of chapter questions

1 5 N.
2 They are opposite.
3 Accelerates in the direction of the force.
4 It decelerates.
5 It accelerates in the direction of the resultant force.
6 140 N.
7 The distance travelled between the driver seeing something and reacting to it by putting on the brakes.
8 Driver tired or under the influence of alcohol or drugs.
9 Weight.
10 Zero.

Chapter 3

Pre Test

1 Energy transferred.
2 Work done = force × distance moved in the direction of the force.
3 Kinetic energy.
4 Energy stored in an elastic object when it is stretched or squashed.
5 Momentum = mass × velocity.
6 Always applies when bodies interact, provided no external forces act.
7 0.5 m/s.
8 0.3 m/s.
9 Crumple zone increases the time for the momentum to become zero, so force on the car decreases.
10 16 000 kg m/s.

Check yourself

3.1
1 Joule.
2 When it moves through a distance.
3 2100 J.

3.2
1 Kinetic energy increases.
2 Joule.
3 50 000 J.

3.3

1 When they are moving.
2 Kg m/s or Ns.
3 10 000 kg m/s.

3.4

1 Zero.
2 They will be equal and opposite.
3 The student with twice the mass has half the speed.

3.5

1 Change in momentum = force × time taken for the change.
2 2500 N.
3 The change in momentum is the same whatever she lands on. On the mat the time taken for the change is increased so the force on her is decreased.

End of chapter questions

1 Joule.
2 400 N.
3 Kinetic energy = $\frac{1}{2}$ mass × speed^2.
4 80 kg.
5 40 000 kg m/s.
6 0.6 m/s.
7 Total momentum before the collision.
8 Velocity has direction.
9 2.5 N.
10 16 000 kg m/s.

EXAMINATION-STYLE QUESTIONS

1 (a) (i) Weight. (1 mark)
(ii) Drag or air resistance or air friction. (1 mark)
(b) Flying at a constant forward speed, (1 mark)
and a constant height. (1 mark)
(c) (i) Lift. (1 mark)
(ii) Thrust. (1 mark)

2 (a) weight = 50 kg × 10 N/kg (1 mark)
weight = 500 N (1 mark)
(b) work done = 200 N × 4.0 m (1 mark)
work done = 800 J (1 mark)

3 (a) (i) change in momentum = force × time (1 mark)
(ii) change in momentum = 200 N × 0.02 s
change in momentum = 4 Ns (1 mark)
change in momentum = mass × change in velocity (1 mark)
velocity = 4 Ns/0.16 kg (1 mark)
velocity = 25 m/s (1 mark)
(b) kinetic energy = $\frac{1}{2}mv^2$ (1 mark)
kinetic energy = $\frac{1}{2}$ × 0.16 kg × 25^2 (1 mark)
kinetic energy = 50 J (1 mark)

4 (a) Total momentum before event equals total momentum after event (1 mark)
(change in momentum is zero scores 2 marks)
(b) collisions (1 mark)
explosions (1 mark)
(c) momentum before = 0.2 kg × 5 m/s + 0.3 kg × 2 m/s (1 mark)
momentum before = 1.6 kg m/s (1 mark)
momentum after = 0.5 kg m/s × v m/s (1 mark)
v = 1.6 kg m/s / 0.5 kg
v = 3.2 m/s (1 mark)

5 (a) Thinking distance is proportional to speed. (1 mark)
Braking distance increases more quickly as speed increases. (1 mark)
(b) stopping distance = thinking distance + braking distance (1 mark)
thinking distance = 17 m (1 mark)
braking distance = 50 m (1 mark)
stopping distance = 67 m (1 mark)
(c) (i) thinking distance same (1 mark)
braking distance increases (1 mark)
(ii) thinking distance increases (1 mark)
braking distance same (1 mark)

Chapter 4

Pre Test

1 Electrons are transferred onto it.
2 They repel each other.
3 Charge moves through it easily.
4 They have free electrons.
5 They pass through a charged grid.
6 They stick to plates, on the walls, with the opposite charge.

Check yourself

4.1

1 Negative.
2 Electrons are rubbed off it.
3 Repulsive.

4.2

1 They do not have free charge carriers.
2 Isolate from the earth.
3 Electric current is the rate of flow of charge.

4.3

1 It prevents static charge building up on the pipe.
2 The paint droplets become positively charged as they leave the sprayer so they repel each other and spread out to form a cloud.
3 By friction with the walls of the pipe.

End of chapter questions

1 The positive charges are on the proton in the nucleus.
2 Friction against your body charges the clothing.
3 Copper or another metal.
4 Connect it to earth.
5 Electrostatic paint sprayer, photocopier, electrostatic smoke precipitator.
6 You would probably get a shock as the charge flowed through you to the earth.

Chapter 5

Pre Test

1 Two or more cells.
2
3 Potential difference = current × resistance.
4 Ampere, A.
5
6 Resistance decreases.
7 Current = $\frac{\text{potential difference of the supply}}{\text{total resistance of the circuit}}$
8 Charge has no choice of route, it must go through every component.
9 Each component is connected across the supply p.d.
10 Add the currents in the individual branches of the circuit.

Check yourself

5.1

1
2
3

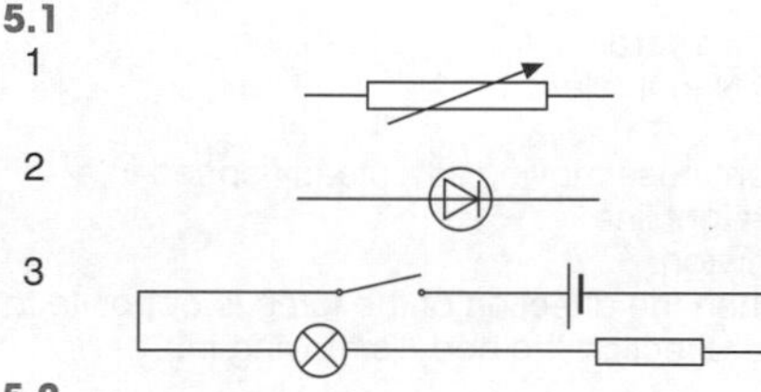

5.2

1 Ammeter.
2 Ohm, Ω.
3 Opposition to current flow.

5.3

1 Resistance increases.
2 No effect.
3 There is current in one direction, in the other the resistance is very high and the current is zero.

5.4

1 Charge stops flowing through any component.
2 Add the individual resistances.
3 p.d. increases.

5.5

1 Little effect on the other components.
2 All the same.
3 Add the currents in the individual branches.

End of chapter questions

1 Diagram using symbols to show components in a circuit.
2

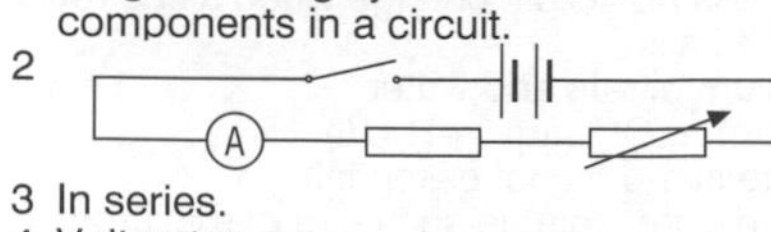

3 In series.
4 Voltmeter.
5
6 As current increases the temperature of the filament increases and its resistance increases.
7 18 Ω.
8 0.5 A.
9 p.d. stays the same.
10 Current decreases.

Chapter 6

Pre Test

1 Current that continuously changes direction.
2 50 Hz.
3 Brown.
4 Plastic or rubber.
5 A thin piece of wire.
6 The case will become live.
7 Power = energy/time.
8 Watt, W.
9 Charge = current × time.
10 Electrical energy to heat.

Check yourself

6.1

1 Current in one direction only.
2 Zero.
3 230 V.

6.2

1 To prevent the case giving someone a shock should it become live.
2 Blue.
3 Brass is a good conductor, does not corrode and is hard.

6.3

1 An electromagnetic switch that will disconnect the supply if the current exceeds a certain value.
2 A large current will flow to earth and melt the fuse.
3 Plastic is an insulator, so current cannot flow through to give a shock to anyone touching the case.

6.4

1 2300 W.
2 90 000 J.
3 3 amp fuse.

6.5

1 Coulomb, C.
2 46 000 J.
3 240 C.

End of chapter questions

1 The direction of the current reverses 50 times each second.
2 d.c. flows in one direction only, a.c. continuously changes direction.
3 Tightly over the outer plastic casing.
4 Green and yellow.
5 Live.
6 It must be higher so the fuse does not melt during normal use. It should be only slightly higher or the current may become large enough to damage the appliance before the fuse melts.
7 Power = current × potential difference.
8 10 A.
9 Energy transformed = potential difference × current.
10 Charge.

Chapter 7

Pre Test

1 The number of protons in the nucleus.
2 The number of protons and neutrons in the nucleus.
3 The atom is a blob of positive matter with negative charges stuck in it.
4 Thin gold foil.
5 The splitting of a nucleus.
6 There are 238 protons and neutrons in the nucleus.
7 The joining together of two nuclei.
8 They all have a positive charge and like charges repel.

Check yourself

7.1
1 Cosmic rays from space, rocks, nuclear power stations, etc.
2 It stays the same.
3 Goes down by two.

7.2
1 Alpha particles were fired at a piece of thin gold foil.
2 Atoms consist mostly of empty space.
3 It is where the mass of the atom is concentrated and is positively charged.

7.3
1 Uranium containing a higher percentage of uranium 235 than occurs naturally.
2 A uranium or plutonium nucleus absorbs a neutron.
3 Uranium 235 and plutonium 239.

7.4
1 Fusion.
2 They must be raised to a very high temperature.
3 By a magnetic field.

End of chapter questions

1 Mass number goes down by four.
2 Atomic number goes up by one.
3 Alpha particles.
4 Empty space.
5 An isotope that will undergo fission if it absorbs a neutron.
6 When a fission occurs, two or three neutrons are produced. If these go on to cause fissions that produce more neutrons that go on to produce more fissions, a chain reaction occurs that gets bigger and bigger.
7 In stars.
8 All nuclei have a positive charge so they repel each other.

EXAMINATION-STYLE QUESTIONS

1 (a) (i) 1.5 V (1 mark)
(ii) 1.5 V (1 mark)
(b) (i) current = p.d./resistance (1 mark)
current = 1. 5 V/3 Ω
current = 0.5 A (1 mark)
(ii) current = 1.5 V/6 Ω
current = 0.25 A (1 mark)
(c) current = 0.5 A + 0.25 A (1 mark)
current = 0.75 A (1 mark)

2 (a) current = power/potential difference (1 mark)
current = 2300/230 V (1 mark)
current = 10 A (1 mark)
(b) 13 A (1 mark)
(c) time = charge/current (1 mark)
time = 2400 C/10 A (1 mark)
time = 240 s (1 mark)
time = 4 minutes (1 mark)

3 (a) Y and Z. (1 mark)
(b) X and Y. (1 mark)
(c) X and Y. (1 mark)
(d) It must lose an electron. (1 mark)
(e) Beta decay – 1 more proton, (1 mark)
1 less neutron. (1 mark)

4 (a) correct symbols for cell, resistor and variable resistor (3 marks)
all connected in series (1 mark)

(b) (i) in series (1 mark)
(ii) in parallel (1 mark)
(c) Straight line with positive slope (1 mark)
passing through the origin (1 mark)

5 Neutron is absorbed by a large nucleus (1 mark)
Large nucleus splits into 2 or 3 smaller nuclei (1 mark)
Plus more neutrons (1 mark)
Releasing energy (1 mark)
Neutrons go on to be absorbed by other nuclei (1 mark)
This is a chain reaction (1 mark)

C2 Data Sheet

1 Reactivity Series of Metals

Potassium	most reactive
Sodium	↑
Calcium	
Magnesium	
Aluminium	
Carbon	
Zinc	
Iron	
Tin	
Lead	
Hydrogen	
Copper	
Silver	
Gold	↓
Platinum	least reactive

(elements in italics, though non-metals, have been included for comparison)

2 Formulae of Some Common Ions

Positive ions		**Negative irons**	
Name	Formula	Name	Formula
Hydrogen	H^{+}	Chloride	Cl^{-}
Sodium	Na^{+}	Bromide	Br^{-}
Silver	Ag^{+}	Fluoride	F^{-}
Potassium	K^{+}	Iodide	I^{-}
Lithium	Li^{+}	Hydroxide	OH^{-}
Ammonium	NH_4^{+}	Nitrate	NO_3^{-}
Barium	Ba^{2+}	Oxide	O^{2-}
Calcium	Ca^{2+}	Sulfide	S^{2-}
Copper(II)	Cu^{2+}	Sulfate	SO_4^{2-}
Magnesium	Mg^{2+}	Carbonate	CO_3^{2-}
Zinc	Zn^{2+}		
Lead	Pb^{2+}		
Iron(II)	Fe^{2+}		
Iron(III)	Fe^{3+}		
Aluminium	Al^{3+}		

3. The Periodic Table of Elements

Key

relative atomic mass **atomic symbol** name atomic (proton) number

1	2											3	4	5	6	7	0
							1 **H** hydrogen 1										4 **He** helium 2
7 **Li** lithium 3	9 **Be** beryllium 4											11 **B** boron 5	12 **C** carbon 6	14 **N** nitrogen 7	16 **O** oxygen 8	19 **F** fluorine 9	20 **Ne** neon 10
23 **Na** sodium 11	24 **Mg** magnesium 12											27 **Al** aluminium 13	28 **Si** silicon 14	31 **P** phosphorus 15	32 **S** sulfur 16	35.5 **Cl** chlorine 17	40 **Ar** argon 18
39 **K** potassium 19	40 **Ca** calcium 20	45 **Sc** scandium 21	48 **Ti** titanium 22	51 **V** vanadium 23	52 **Cr** chromium 24	55 **Mn** manganese 25	56 **Fe** iron 26	59 **Co** cobalt 27	59 **Ni** nickel 28	63.5 **Cu** copper 29	65 **Zn** zinc 30	70 **Ga** gallium 31	73 **Ge** germanium 32	75 **As** arsenic 33	79 **Se** selenium 34	80 **Br** bromine 35	84 **Kr** krypton 36
85 **Rb** rubidium 37	88 **Sr** strontium 38	89 **Y** yttrium 39	91 **Zr** zirconium 40	93 **Nb** niobium 41	96 **Mo** molybdenum 42	[98] **Tc** technetium 43	101 **Ru** ruthenium 44	103 **Rh** rhodium 45	106 **Pd** palladium 46	108 **Ag** silver 47	112 **Cd** cadmium 48	115 **In** indium 49	119 **Sn** tin 50	122 **Sb** antimony 51	128 **Te** tellurium 52	127 **I** iodine 53	131 **Xe** xenon 54
133 **Cs** caesium 55	137 **Ba** barium 56	139 **La*** lanthanum 57	178 **Hf** hafnium 72	181 **Ta** tantalum 73	184 **W** tungsten 74	186 **Re** rhenium 75	190 **Os** osmium 76	192 **Ir** iridium 77	195 **Pt** platinum 78	197 **Au** gold 79	201 **Hg** mercury 80	204 **Tl** thallium 81	207 **Pb** lead 82	209 **Bi** bismuth 83	[209] **Po** polonium 84	[210] **At** astatine 85	[222] **Rn** radon 86
[223] **Fr** francium 87	[226] **Ra** radium 88	[227] **Ac*** actinium 89	[261] **Rf** rutherfordium 104	[262] **Db** dubnium 105	[266] **Sg** seaborgium 106	[264] **Bh** bohrium 107	[277] **Hs** hassium 108	[268] **Mt** meitnerium 109	[271] **Ds** darmstadtium 110	[272] **Rg** roentgenium 111	Elements with atomic numbers 112–116 have been reported but not fully authenticated						

* The Lanthanides (atomic numbers 58 –71) and the Actinides (atomic numbers 90–103) have been omitted.

Cu and **Cl** have not been rounded to the nearest whole number.